Immunology:
An Illustrated Outline

Fifth Edition

Immunology:
An Illustrated Outline

Fifth Edition

David Male

NEW YORK AND LONDON

Vice President: Denise Schanck
Development Editor: Monica Toledo
Production Editor: Georgina Lucas
Copy Editor: Bruce Goatly
Proofreader: Karin Henderson
Illustrations: Nigel Orme
Cover Design: Andy Magee

ISBN 978-0-8153-4501-5

Library of Congress Cataloging-in-Publication Data
Male, David K., 1954-
Immunology : an illustrated outline / David Male. -- Fifth edition.
pages cm
Includes bibliographical references and index.
ISBN 978-0-8153-4501-5 (pbk.)
1. Immunology--Outlines, syllabi, etc. 2. Clinical immunology--Outlines, syllabi, etc. I. Title.
QR182.55.M35 2014
616.07'9--dc23

2013018416

Published by Garland Science, Taylor & Francis Group, LLC,
an informa business,
711 Third Avenue, 8th floor, New York, NY 10017, USA,
and 3 Park Square, Milton Park, Abingdon, OX14 4RN, UK.

15 14 13 12 11 10 9 8 7 6 5 4 3 2 1

Visit our website at http://www.garlandscience.com

Preface

This book serves three different functions. It can be used as a dictionary of immunology or as a review guide for undergraduate, graduate, or medical students taking an immunology course. It can also provide a concise overview of basic immunology for readers who have not studied the subject previously but need immunology as a background for undergraduate or postgraduate work. The book also serves as a reference for those who already have a background in immunology but require a refresher on specific content.

Readers who already know some immunology and require a summary of particular aspects should consult the contents pages. The book is divided into five sections, each of which contains a number of related topics that are generally set out on double-page spreads. These topics are arranged in a logical sequence, so that sections 1–3 can be read as a short course of basic immunology, while section 4 provides the basis of clinical immunology and section 5 provides a review of immunological techniques.

To use the book as a dictionary, look up the word or abbreviation in the "Index of Terms." This gives a single page number where a definition of the word will be found; associated terms will be found on the same page. Page references to particular topics set out over several pages are indicated in bold. Items found as entries in tables are given in italics.

This latest edition of the book has been comprehensively revised, to highlight the latest understanding of the subject, particularly in the areas of innate immune defenses and immunological techniques. Naturally a book of this kind cannot include everything of interest to immunologists. I have tried to cover all the essential areas of the subject, but I should be pleased to know when readers consider that particular subjects deserve further detail.

I am most grateful to colleagues who have allowed me to use micrographs and photographic illustrations; their contributions are individually acknowledged in the legends. For this edition, I am very pleased to be working with a new editorial team, including Denise Schanck, Monica Toledo, and Georgina Lucas from Garland Science, and with a new illustrator, Nigel Orme.

Contents

Index of Terms

Note: **Bold** page numbers identify topics set out on several pages; roman page numbers identify entries in the text, including definitions; *italic* page numbers indicate additional information in tables or figures.

D

J

K

L

M

N

O

P

Q

R

U

V

W

X

Immunology
An Illustrated Outline

1 The Immune System

INTRODUCTION

The immune system has evolved to protect the body from damage caused by microorganisms—bacteria, fungi, viruses, and parasites. This defensive function is performed by leukocytes (white blood cells) and a number of accessory cells, which are distributed throughout the body but are found particularly in lymphoid organs, including the bone marrow, thymus, spleen, and mucosa-associated lymphoid tissues (MALT). Lymphoid organs are strategically placed to protect different areas of the body from infection. Cells migrate between these tissues via the bloodstream and lymphatic system. As they do so they interact with each other to generate coordinated immune responses, aimed at eliminating pathogens or minimizing the damage they cause.

Lymphocytes are key cells controlling the immune response. They do so by recognizing molecules produced by pathogens. They can also recognize molecules on the cells of the body, although they do not normally react against the body's own tissues. Molecules recognized by lymphocytes are referred to as 'antigens.' Lymphocytes are of two main types: B cells, which produce antibodies, and T cells, which have a number of functions, including 1. helping B cells to make antibody; 2. recognizing and destroying cells that have become infected with intracellular pathogens; 3. activating phagocytes to destroy pathogens that they have taken up, and 4. regulating the level and quality of the immune response. Lymphocytes recognize foreign material by specific cell-surface antigen receptors. To recognize the enormous variety of different molecules, the antigen receptors must be equally diverse. Each lymphocyte makes only one type of antigen receptor and thus can only recognize a very limited number of antigens, but as the receptors differ on each clone of cells, the lymphocyte population, as a whole, can recognize a vast range of different antigens. A third, minor, population of lymphocytes (NK cells) also contributes to antiviral defenses.

Phagocytes include blood monocytes, macrophages, and neutrophils. They can internalize (phagocytose) pathogens, antigens, and cell debris and break them down. Antibodies and various other recognition molecules bound to the pathogens facilitate this process. Macrophages can also process and present antigens, so that they can be recognized by T cells.

Accessory cells include eosinophil and basophil granulocytes, mast cells, platelets, and antigen-presenting cells (APCs). Eosinophils have a role in protection against some parasites. Basophils, mast cells, and platelets contain a variety of molecules that mediate inflammation. APCs are a functionally defined group of cells; both B cells and macrophages can present antigen, but leukocyte dendritic cells are particularly important in presenting antigen to naive T cells, which have not previously encountered their specific antigen.

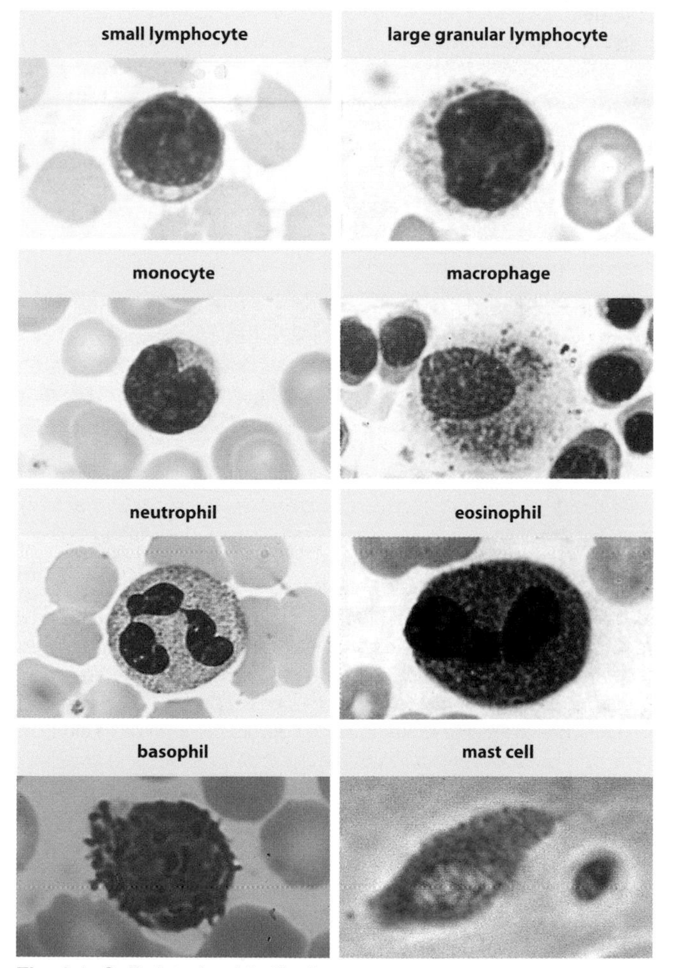

Fig. 1.1 Cells involved in the immune response.
Macrophage courtesy of A. V. Hoffbrand.

LYMPHOCYTES

Lymphocytes constitute about 20% of the total blood leukocytes. The two major populations of lymphocytes, T cells and B cells, are small lymphocytes responsible for recognizing antigens or antigen fragments. A third population of large granular lymphocytes (LGLs) recognizes changes in host cells that may occur when they become infected.

Large granular lymphocytes (LGLs) are morphologically defined cells containing large amounts of cytoplasm, with azurophilic granules, which constitute 5–15% of the blood T cells. Both NK cells and $\gamma\delta$ T cells have this morphology.

T cells are lymphocytes that develop in the thymus. This organ is seeded by lymphocytic stem cells from the bone marrow during embryonic development. The cells then develop their T-cell antigen receptors (TCR) and differentiate into the two major peripheral T cell subsets; the helper T cells express CD4, and the cytotoxic T cells express CD8. T cells can also be differentiated into two populations according to whether they use an $\alpha\beta$ (TCR2) or a $\gamma\delta$ (TCR1) antigen receptor. The essential role of T cells is to recognize antigens associated with cells of the host.

$\gamma\delta$ T cells express the $\gamma\delta$ form of the T cell receptor. They constitute <5% of the total T cells, but they are more common in particular sites, including the gut, skin, and vagina. They branch early from the main thymic developmental pathway and recognize different antigens from $\alpha\beta$ T cells, including carbohydrates and intact proteins.

Intraepithelial lymphocytes (IELs) are mixed populations of cells found in submucosal tissues. 10–40% are $\gamma\delta$ T cells, with a dendritic appearance. The remainder are mostly $CD8^+$ T cells.

Cytotoxic T (Tc) cells are capable of destroying virally infected or allogeneic cells. Most Tc cells express CD8 and recognize antigen associated with MHC class I molecules, which may be expressed on all nucleated cells of the body.

Helper T (TH) cells perform a number of functions, including helping B cells to divide, differentiate, and secrete antibody, activating macrophages to destroy pathogens that they have phagocytosed and recruiting cells to sites of inflammation. The functions are carried out by different subsets of TH cells, which differentiate from a common precursor (TH0) and can be distinguished by the cytokines, which they secrete. The majority of TH cells express CD4 and recognize antigenic peptides presented on the surface of APCs by MHC class II molecules.

TH1, TH2, and TH17 cells are subsets of helper T cells originally identified according to the cytokines they produce. They differentiate from a common precursor (TH0). Differentiation of TH1 cells is promoted by interleukin (IL)-12 and interferon-γ (IFN-γ), TH2 cells by IL-4, and TH17 cells by transforming growth factor-β (TGF-β) and IL-6. Dendritic cells are antigen-presenting cells which are most effective at presenting antigen to naive T cells. TH1 cells can recognize antigen presented by mononuclear phagocytes, and they interact with these cells by releasing IFN-γ, which acts as a macrophage activation factor. TH2 cells release cytokines, such as IL-4 and IL-5, which are required for B-cell development into plasma cells. Both TH1 and TH2 cells can modulate the antibody response by affecting the classes of immunoglobulin produced. TH17 cells release cytokines that promote inflammatory responses, particularly by acting on neutrophils. Some cell-surface markers are preferentially expressed on a subset of the helper T cells. For example, the chemokine receptors CCR5 and CXCR3 are more prevalent on TH1 cells, whereas CCR3 and CCR4 are higher on TH2 cells. All helper T cells can promote the development and activation of cytotoxic T cells and NK cells, which recognize and kill infected target cells.

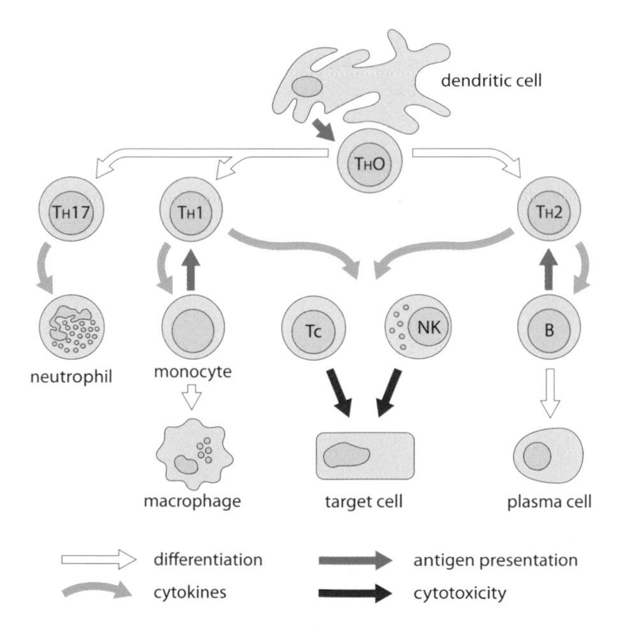

Fig. 1.2 Lymphocyte interactions.

Regulatory T cells (TREG) are a population of T cells usually identified by the expression of the transcription factor Foxp3 and/or high expression of the IL-2 receptor, CD25. They can be either $CD4^+$ or $CD8^+$ and are important in controlling secondary immune responses and inflammation, particularly in the gut. They also limit some autoimmune and hypersensitivity reactions, acting by direct cell–cell interactions, or by the release of anti-inflammatory cytokines including IL-10, IL-35, and TGF-β. They can also inhibit activation of other T cells by mopping up IL-2, which is required for T-cell proliferation.

B cells are lymphocytes that develop in the fetal liver and subsequently in bone marrow. In birds, they develop in a specialized organ, the bursa of Fabricius. Mature B cells express surface immunoglobulin, which acts as the B-cell antigen receptor (BCR). They are distributed throughout the secondary lymphoid tissues, particularly in the follicles of lymph nodes, spleen, and Peyer's patches. They respond to antigenic stimuli by dividing and differentiating into plasma cells.

Plasma cells/Antibody-forming cells (AFC) are terminally differentiated B cells, with expanded cytoplasm containing arrays of rough endoplasmic reticulum, devoted to the synthesis of secreted antibody. Plasma cells are seen in the red pulp of spleen, the medulla of lymph nodes, the MALT, and occasionally in sites of inflammation.

B1 and B2 cells are B cell subsets. In adults most B cells are of the B2 subset. They generate a wide range of antigen receptors, mature in germinal centers and respond well to T-dependent antigens and co-stimulation via CD40. The B1 subset was originally distinguished by the phenotype $CD5^+$, $CD43^+$, $CD23^-$. B1 cells develop early, have a more limited range of receptors than B2 cells, respond to a number of common microbial antigens, and sometimes produce autoantibodies. They are absent from lymph nodes, constitute 5% of splenic B cells, and are important in mucosal immunity.

Naive/Virgin lymphocytes are cells that have not encountered their specific antigen. They express high-molecular-weight variants of the leukocyte common antigen (for example, CD45RA).

Clonal selection describes the way in which lymphocytes are activated. During development each lymphocyte generates an antigen receptor with a single antigen specificity. If antigen is encountered, only the few clones of lymphocytes that can recognize it are stimulated to divide to provide a large pool of effector and memory cells. This is referred to as 'clonal selection.'

Memory cells are populations of long-lived T or B cells with some capacity for self-renewal that have been previously stimulated with antigen and can make an accelerated response if they encounter it again. Memory B cells carry IgG or IgA as their antigen receptor, which is of higher affinity than the receptor (IgM or IgD) on naive cells. Memory T cells express the CD45RO variant of leukocyte common antigen as well as increased levels of adhesion molecules LFA-3 and VLA-4. Immunological memory depends on both the production of memory cells and the increase in the numbers of antigen-specific cells produced in the primary response.

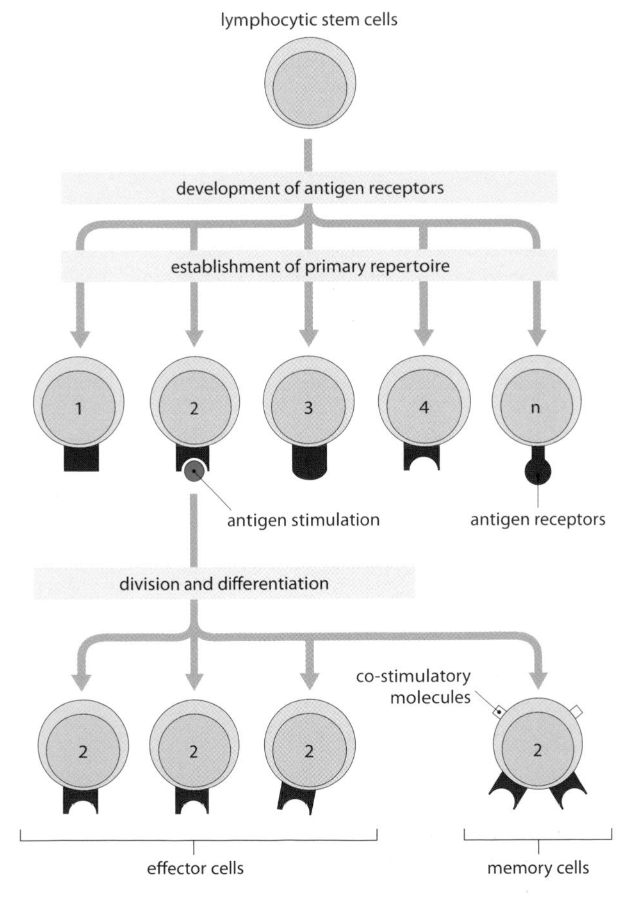

Fig. 1.3 Clonal selection and lymphocyte development.

NK CELLS

Null cells/Third population cells are all descriptions of a distinct population of lymphocytes constituting 5–15% of blood mononuclear cells. They lack conventional antigen receptors (TCR, BCR), but express some markers of both T cells and mononuclear phagocytes. The great majority of these cells have the appearance of large granular lymphocytes (LGLs). Functionally they act as natural killer (NK) cells.

NK (Natural killer) cells are capable of killing a variety of virally infected or transformed target cells, particularly cells that have lost or reduced expression of MHC class I molecules, or express allogeneic MHC molecules. Thus they provide a second line of defense against viruses that attempt to evade immune recognition by downregulating the expression of MHC molecules. NK cells can use various receptors including the Fc receptor (CD16), CD2, CD69, KIRs, and lectin-like receptors to recognize target cells. NK cells kill their targets using similar mechanisms to Tc cells, with the granule components perforin and granzymes being most important.

Killer immunoglobulin-like receptors (KIR) are a family of receptors that bind to MHC class I molecules and are used by NK cells to recognize their targets. KIRs may have two or three extracellular immunoglobulin-like domains and they are produced in two forms, an inhibitory receptor with a long cytoplasmic tail containing ITIM motifs, and an activating receptor with a short cytoplasmic tail that can interact with ITAM-containing adaptor molecules. As MHC molecules have diversified, so have the KIRs that recognize them. There are several gene loci encoding KIRs (CD158). The number varies between individuals and there is much polymorphism. Each NK cell expresses a subset of the available NK-cell receptors and can therefore recognize loss or change in one group of MHC molecules. T cells may also express some KIRs after activation by antigen. NK cells respond to a balance of signals from the activating and inhibitory receptors in order to interact with cells of the body and respond to changes in their MHC expression.

Lectin-like receptors are a family of receptors consisting of two polypeptides, NKG2 and CD94, which are present on most NK cells and on some cytotoxic T cells. They recognize leader peptides of MHC molecules, presented by a nonclassical MHC-encoded molecule, HLA-E. Loss of MHC molecules by a cell leads to a global reduction of MHC peptides presented by HLA-E. Like KIRs the lectin-like receptors are produced in inhibitory and activating forms depending on their cytoplasmic tail.

MARKERS

CD system. Leukocytes are differentiated by their cell-surface molecules, which are identified by monoclonal antibodies. The most readily accessible marker of lymphocytes is their antigen receptor. The B-cell receptor (BCR) is surface immunoglobulin, whereas T cells express the TCR. Most other markers are designated according to the CD system of nomenclature. Some of these markers are specific for individual populations of cells or particular phases of cellular differentiation. Others appear only on activated or dividing cells. Many CD markers are present at varying levels on several different cell types, so that each subset of lymphocytes expresses a unique profile of surface markers. More than 300 individual molecules are recognized in the CD system and some of them are found on cells other than leukocytes. The table overleaf gives the identity and cellular distribution of the more important CD molecules. Particularly important are the markers used to distinguish T cells (CD2, CD3), the main T-cell subsets (CD4, CD8), activated T cells and TREG cells (CD25), B cells (CD19, CD20, CD79), mononuclear phagocytes (CD64, CD68), and NK cells (CD56).

Families of proteins Despite the enormous number of cell-surface molecules, most of them belong to just a few families, which share common structural features. Such families include the immunoglobulin supergene family (IgSF), the 4-transmembrane (tm4) and 7-transmembrane (tm7) families, the C-type lectins, the integrins, and the complement control proteins (CCP) (Fig. 1.4).

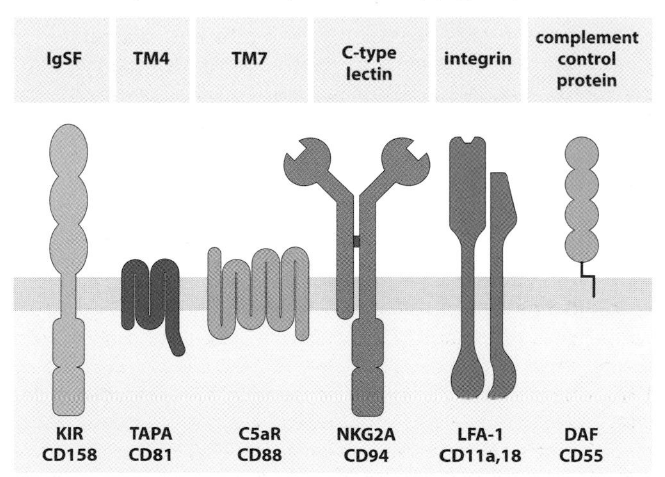

Fig. 1.4 Families of cell-surface molecules, with examples.

	identity/function	T cell	B cell	NK cell	monocyte/ macrophage	granulocyte	others
CD1	presentation of lipoprotein antigens	Thy					IDC
CD2	binds CD58 or CD48; co-stimulation						
CD3	TCR; signal transduction						
CD4	MHC class II receptor						
CD5	differentiates B-cell subset						
CD8	MHC class I receptor						
CD11a	LFA-1; integrin α chain						
CD11b	CR3 (Mac-1); integrin α chain						
CD11c	CR4; integrin α chain						
CD13	aminopeptidase N						
CD14	receptor for lipopolysaccharide						
CD15	Lewis^{X}/sialyl Le^{X}; binds E-selectin						
CD16	FcγRIII						
CD18	β_2 integrin (see CD11)						
CD19	B-cell co-receptor complex (see CD21 & 81)						
CD20	B-cell regulation						
CD21	CR2; B-cell co-receptor complex						FDC
CD23	FcεRII				★	Eo	
CD25	IL-2R α chain	★	★		★		
CD28	binds CD80 and CD86; co-stimulation		★				
CD29	β_1 integrin (see CD49)						
CD30	regulates proliferation and cell death	★	★				
CD31	PECAM; regulates adhesion						End
CD32	FcγRII						
CD34	cell adhesion						End
CD35	CR1						FDC
CD37	signal transduction						
CD38	ADP ribosyl cyclase, regulates proliferation	★	PC				
CD40	binds CD154; co-stimulation						IDC
CD43	leukosialin						
CD44	matrix adhesion						
CD45	leukocyte common antigen (LCA)						
CD45R	restricted LCA						
CD46	membrane cofactor protein (MCP)						
CD48	binds CD2 (mouse)						
CD49a	VLA-1; integrin α chain	★					
CD49b	VLA-2; integrin α chain	★					
CD49c	VLA-3; integrin α chain						
CD49d	VLA-4; binds VCAM-1 & fibronectin	★					
CD49e	VLA-5; integrin binds fibronectin						
CD49f	VLA-6; integrin binds laminin						
CD50	ICAM-3; co-stimulation						

Key: Useful marker; Subpopulation; ★ Activated cells

B = Basophil End = Endothelium Eo = Eosinophil FDC = Follicular dendritic cell

IDC = Interdigitating dendritic cell Thy = Thymocytes PC = Plasma cell DC = dendritic cell

Fig. 1.5 CD Markers.

	identity/function	T cell	B cell	NK cell	monocyte/ macrophage	granulocyte	others
CD53	signal transduction						
CD54	ICAM-1; adhesion	★	★	★			End
CD55	DAF						
CD56	NCAM; adhesion	★	★				
CD57	HNK-1						
CD58	LFA-3; co-stimulation						
CD59	protectin						
CD62E	E-selectin						End
CD62P	P-selectin						End
CD62L	L-selectin						
CD64	FcγRI						
CD68	macrosialin						
CD71	transferrin receptor	★	★	★	★		★
CD73	ecto 5′-nucleotidase						
CD74	MHC class II-associated chain						IDC
CD79ab	sIg; signal transduction						
CD80	binds CD28; co-stimulation						
CD81	TAPA; B cell co-receptor complex						
CD85	inhibits T-cell/NK-cell cytotoxicity						DC
CD86	binds CD28; co-stimulation		★				
CD87	urokinase plasminogen activator receptor	★					
CD88	C5a receptor						
CD89	FcαR						
CD90	Thy-1						Thy
CD94	inhibits NK cell cytotoxicity (see CD159a)						
CD95	binds CD178; cytotoxicity						
CD102	ICAM-2						End
CD103	integrin α chain; intra-epithelial adhesion						
CD105	endoglin; regulates TGF-β receptor						End
CD106	VCAM-1						End
CD143	angiotensin-converting enzyme						End
CD144	VE-cadherin; homotypic adhesion						End
CD152	CTLA-4, binds CD80/86; inhibits activation	★					
CD153	binds CD30	★					
CD154	binds CD40	★				B Eo	
CD158	killer inhibitory receptor family						
CD159a	inhibits NK cytotoxicity (see CD94)						
CD162	PSGL-1, adhesion						
CD178	Fas ligand, binds CD95	★					
CD200	inhibits immune response	★					End
CD204	macrophage scavenger receptor						
CD206	macrophage mannose receptor						IDC
CD244	receptor for CD48, NK cell adhesion						
CD247	T-cell receptor ζ chain						
CD273	PD-1 receptor						DC
CD305	inhibitory receptor, LAIR-1						

ANTIGEN-PRESENTING CELLS

Antigen-presenting cells (APC) are functionally defined cells that take up antigens and present them to lymphocytes in a form they can recognize. Some antigens are taken up by APCs in the periphery and transported to the secondary lymphoid tissues, whereas other APCs are resident in these tissues and intercept antigen as it arrives. B cells recognize antigen in its native form, but T_H cells recognize antigenic peptides that have become associated with MHC molecules. Consequently, in order to present antigen to a T_H cell, an APC must internalize it, process it into fragments and re-express it at the cell surface in association with MHC class II molecules. In addition, many APCs provide co-stimulatory signals to lymphocytes, either by direct cellular interactions or by cytokines. Dendritic cells, macrophages, B cells, and sometimes tissue cells can present antigen to T_H cells.

Dendritic cells (DCs) are a distinct set of APCs distributed in many tissues of the body, which differentiate from either lymphoid or myeloid precursors. Dendritic cells expressing MHC class II molecules migrate to lymph nodes via the lymphoid system carrying antigen, and there they upregulate co-stimulatory molecules required for T-cell activation (CD40, CD80, CD86). In lymph nodes they appear as interdigitating dendritic cells (IDCs) in the paracortex and they are very effective at presenting antigen to naive $CD4^+$ T cells. A minor population of dendritic cells enters the lymphoid tissues directly from the blood.

Langerhans cells (Veiled cells) are myeloid dendritic cells of the skin that pick up antigen and transport it to regional lymph nodes. They express CD207 (langerin), CD1, and high levels of MHC class II molecules and they have a characteristic racket-shaped granule, the Birbeck granule (function unknown). In afferent lymph they are seen as veiled cells and in lymph nodes they develop into dendritic cells. They are particularly important in the development of contact hypersensitivity; skin-sensitizing agents and UV radiation induce their emigration from the skin.

Macrophage APCs Macrophages phagocytose antigen and some of them can also process and present it. MHC class II and co-stimulatory molecules (B7.1/B7.2, CD80/86) are induced by microbial molecules acting on Toll-like receptors, allowing the macrophages to present antigen effectively to T_H1 cells. Activated macrophages also upregulate adhesion molecules (such as ICAM-1) and secrete IL-1. The recirculating macrophages of secondary lymphoid tissues are mostly seen in the medulla of lymph nodes and the red pulp of spleen.

Follicular dendritic cells (FDCs) are present in follicles of spleen and lymph nodes, where they are tightly surrounded by lymphocytes. Complement-fixing immune complexes localize on the surface of these cells via Fc and C3 receptors, where they are presented mainly to B cells. This form of complex localization and presentation is important in the development of B-cell memory.

Iccosomes are beaded cytoplasmic structures present on filopodia of FDCs, which are thought to act as a long-term repository of antigens. They bud off and may be internalized by B cells.

Marginal zone macrophages are present in the marginal zone of the splenic periarteriolar lymphatic sheath and along the marginal sinus of lymph nodes. T-independent antigens such as polysaccharides, tend to localize on these cells, where they are often very persistent. Marginal zone macrophages express sialoadhesin (siglec-1, CD169), a lectin-like receptor for glycoconjugates, and they present antigens primarily to B cells.

Facultative APCs Many cells of the body can be induced to express MHC class II when stimulated by IFN-γ. Sometimes they can present antigen to $CD4^+$ T cells, although they often fail to induce T cell division, owing to their inability to deliver co-stimulatory signals.

APC	location	MHC class II	co-stimulatory molecules	present to:
interdigitating dendritic cell	lymph node paracortex	++	B7.1 B7.2 ICAM-1 ICAM-3	naive T cell
B cell	germinal center	+ → ++	B7.1 B7.2 ICAM-1 inducible	T cell
macrophage	tissues and lymphoid organs	0 → ++	B7 inducible ICAM-3 ICAM-1 inducible	T cell
marginal zone macrophage	marginal zone of spleen and lymph node	–	–	T-ind ags → B cell
follicular dendritic cell	germinal center	–	iccosome components (such as C3b)	B cell

Fig. 1.6 Antigen-presenting cells.

PHAGOCYTES AND AUXILIARY CELLS

Mononuclear phagocyte system is the collective term for the long-lived phagocytic cells distributed throughout the organs of the body. They are derived from bone marrow stem cells and express receptors for immunoglobulin (FcγR) and complement (CR1, CR3, and often CR4). They phagocytose antigenic particles and some have the ability to present antigen to lymphocytes. This group includes:

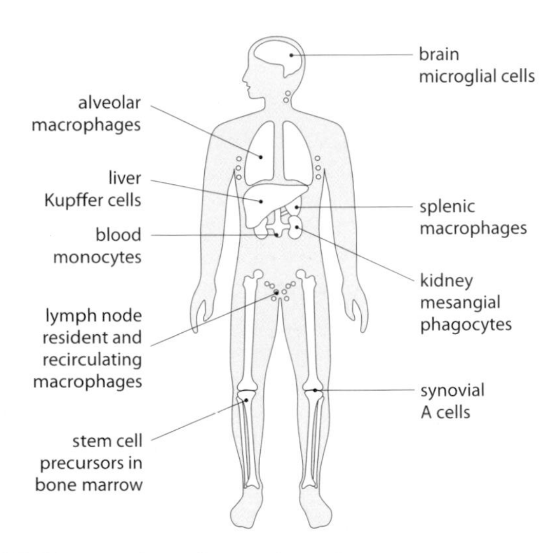

Fig. 1.7 Mononuclear phagocyte system.

Monocytes are circulating cells that constitute ~5% of total blood leukocytes, which can migrate into tissues and differentiate into macrophages. They have a horseshoe-shaped nucleus, azurophilic granules, and many lysosomes.

Macrophages are large phagocytic cells found in most tissues and lining serous cavities and the lung. Resident macrophages may remain in tissues for years, while others recirculate through secondary lymphoid tissues, where they may function as APCs. The development of macrophages is promoted by macrophage colony stimulating factor (M-CSF). They differentiate into distinct subpopulations in different locations, which is influenced by signals from cells of the tissue.

Kupffer cells are phagocytes that lie along the liver sinusoids. Much of the antigen entering the body through the gut is removed by these cells.

Mesangial phagocytes line the glomerular endothelium where the capillaries enter the Bowman's capsule.

Microglial cells are resident phagocytes of the brain, with a distinctive dendritic morphology. Colonization occurs primarily before birth and in the neonatal period.

Synovial A cells are phagocytes that lie on the synovial membrane of the joints, in contact with synovial fluid.

Granulocytes (polymorphs), recognizable by their multi-lobed nuclei and numerous cytoplasmic granules, constitute the majority of blood leukocytes. They are classified by staining as:

Neutrophils—professional phagocytes and the most abundant of the blood leukocytes (>70%). They spend less than 48 hours in the circulation before migrating into the tissues under the influence of chemotactic stimuli, where they phagocytose material and eventually die. They have receptors for antibody and complement to facilitate the uptake of opsonized particles.

Eosinophils—comprising 2–5% of blood leukocytes. Their granules contain a crystalloid core of basic protein, which can be released by exocytosis, causing damage to a number of pathogens, particularly parasitic worms. The granules also contain histaminase and aryl sulfatase, which downregulate inflammatory reactions.

Basophils—constituting <0.5% of blood leukocytes. Their granules contain inflammatory mediators and they are in some ways functionally similar to mast cells.

Mast cells are present in most tissues, adjoining the blood vessels. They contain numerous granules with inflammatory mediators, such as histamine and platelet-activating factor (PAF), released by triggering with C3a or C5a, or by cross-linking of surface IgE bound to their high-affinity IgE receptor (FcεRI). Stimulation causes them to produce prostaglandins and leukotrienes. There are two types of mast cell, thought to be derived from a common precursor.

Connective tissue mast cells (CTMCs)—the main tissue-fixed mast cell population. They are ubiquitous, distributed around blood vessels, and contain large amounts of histamine and heparin. They are inhibited by sodium cromoglycate.

Mucosal mast cells (MMCs)—present in the gut and lung. They are dependent on IL-3 and IL-4 for their proliferation and they are increased during parasitic infections of the gut.

LYMPHOID SYSTEM

Primary and **Secondary lymphoid tissue** Lymphocytes are derived from bone marrow stem cells and initially develop in the primary lymphoid tissues, T cells in the thymus, and B cells in the bone marrow. Mature cells expressing antigen receptors seed the secondary lymphoid tissues, the spleen, lymph nodes, and collections of mucosa-associated lymphoid tissues (MALT).

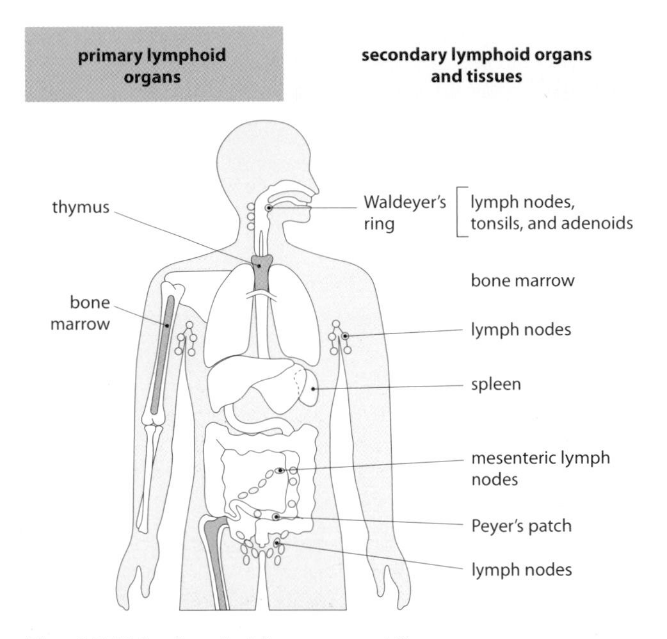

Fig. 1.8 Major lymphoid organs and tissues.

Lymphocyte traffic Lymphocytes leave the circulation by traversing specialized venules (HEV) in the lymph nodes and MALT. They recirculate via the lymphatic system, through chains of lymph nodes, back to the circulation. Recirculation gives lymphocytes the opportunity to contact their antigen.

High endothelial venules (HEV) are present in most secondary lymphoid tissues and may be induced in other tissues during severe persistent immune reactions. They are lined with distinctive columnar cells expressing site-specific sets of glycosylated adhesion molecules and chemokines (such as CCL21). Up to 25% of lymphocytes, including naive cells, passing through the lymphoid tissues, bind to the adhesion molecules and migrate across HEV.

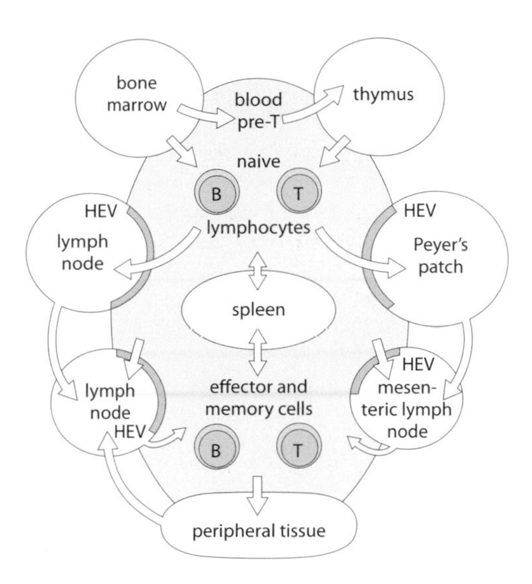

Fig. 1.9 Lymphocyte traffic.

Lymphatic system is a system of vessels covering the entire body that is responsible for draining tissues and returning the transudate to the blood. It also acts as a route for the movement of antigens from the periphery to the lymph nodes and for the recirculation of lymphocytes and dendritic cells.

Thoracic duct and Right lymphatic duct are the main lymphatic vessels draining into the blood. Recirculating cells from the trunk, internal organs, and lower limbs pass through the thoracic duct into the subclavian vein. The right lymphatic duct drains the upper right quadrant of the body.

Mucosa-associated lymphoid tissue (MALT) is a general term for the unencapsulated lymphoid tissues that are seen in submucosal areas of the respiratory, gastrointestinal, and urogenital systems. These protect potential sites of pathogen invasion. The majority of the body's lymphocytes are found in the MALT.

Bronchus-associated lymphoid tissue (BALT) is the name for the part of the MALT associated with the respiratory system.

Waldeyer's ring is the lymphoid tissue of the neck and pharynx, which includes the adenoids, tonsils, and regional lymph nodes.

Tonsils and **Adenoids** are pharyngeal parts of the MALT that are particularly rich in B cells, arranged in lymphoid follicles.

LEUKOCYTE DEVELOPMENT

Bone marrow is a hemopoietic tissue present in long bones and the axial skeleton. A network of venous sinuses is arranged around a central artery and vein, and these permeate the developing cells. All blood cells are derived from bone marrow stem cells, and 10% of the marrow cells are lymphocytes, occurring in clusters around the radial arteries. In adult mammals B cells develop and differentiate in the marrow. Stromal cells secrete cytokines, including stem-cell factor (SCF) and IL-7, required for the early development of pre-T and pre-B cells. Small numbers of mature lymphocytes also reside in the bone marrow.

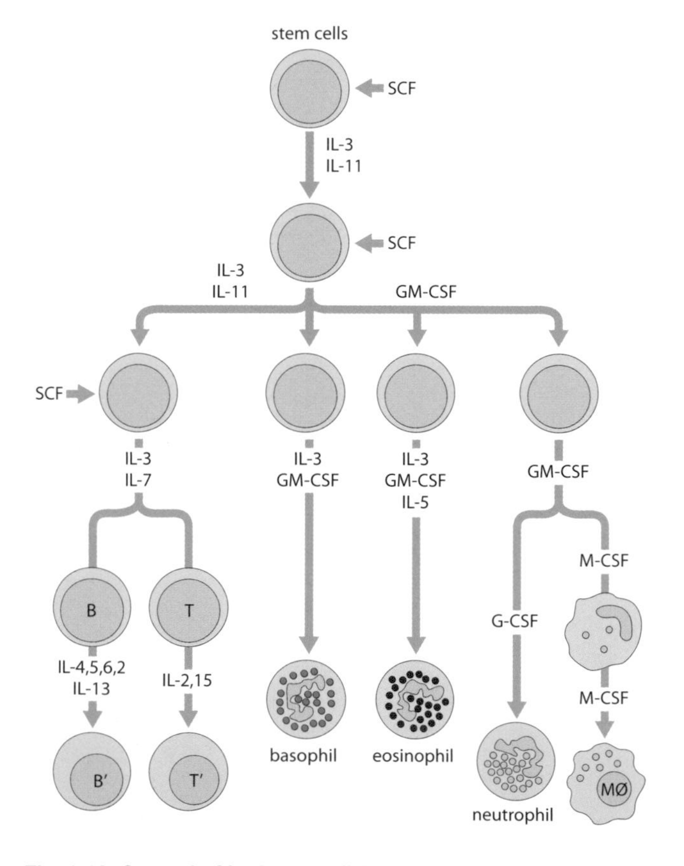

Fig. 1.10 Control of leukocyte differentiation by cytokines.

Stem cell factor (SCF, Steel factor) is a cytokine that acts on a variety of lineages to promote division. Differentiating cells lose their requirement for SCF.

c-Kit (CD117) is the receptor for SCF, present on T and B cell precursors. It disappears when the lymphocyte precursors start to recombine their antigen receptor genes, but is retained on hemopoietic stem cells. A subpopulation of NK cells also expresses c-Kit permanently. Mast-cell precursors also express c-Kit. The receptor also binds mast-cell growth factor (MGF).

Colony-stimulating factors (CSFs) control the differentiation of hemopoietic stem cells, both in the bone marrow and in the periphery (Fig. 1.10). This group of cytokines includes granulocyte, macrophage, and granulocyte/macrophage CSFs (G-CSF, M-CSF, and GM-CSF, respectively), which promote the differentiation of their specific subsets of leukocytes. In addition, IL-3 (pan-specific hemopoietin), IL-5, IL-7, and IL-11 are functional members of this group.

Myeloid cells are the granulocyte and mononuclear phagocyte lineages that develop from a common stem cell. The stem cell (CFU-GM) expresses CD34 (also present on resting endothelium) and MHC class II, which is lost as differentiation proceeds.

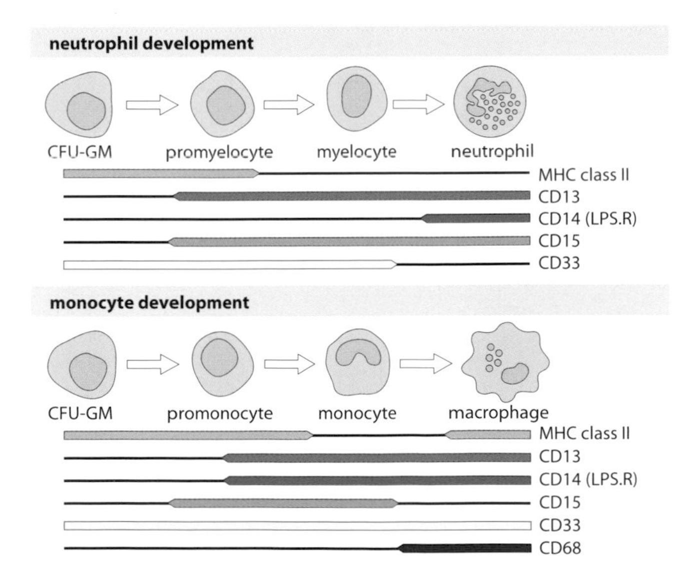

Fig. 1.11 Differentiation of myeloid cells.

THYMUS

The thymus is a lymphoid organ overlying the heart, seeded by lymphoid stem cells from the bone marrow, which differentiate into T cells. It is bilobed and organized into lobules separated by connective tissue septae (trabeculae). Each lobule contains a peripheral cortex and central medulla.

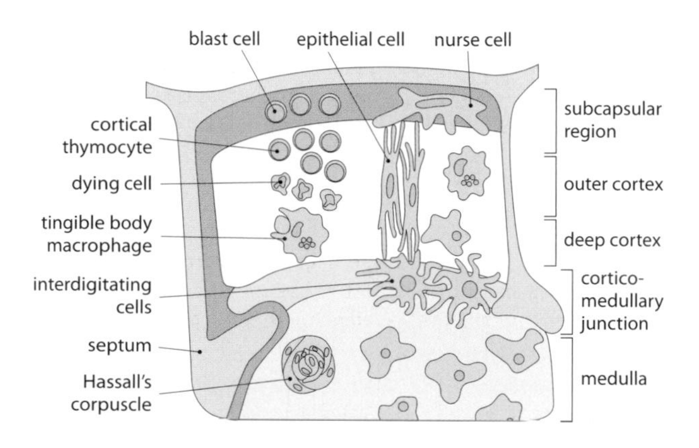

Fig. 1.12 The structure of a thymus lobule.

Thymocytes are thymic lymphocytes. The repertoire of T-cell antigen receptors develops, and the deletion of autoreactive cells occurs during T-cell maturation in the thymus, by interaction with APCs. The process involves proliferation of immature cells, but most thymocytes die during selection by apoptosis.

Thymic cortex. The outer zone of the thymus contains about 85% of the total thymocytes. These cells are immature, express CD1 in humans, and divide rapidly. Most cortical thymocytes express both CD4 and CD8, and are hence called 'double-positive' cells.

Thymic medulla contains relatively few lymphocytes, but they are more mature than those in the cortex. Peripheral cell populations expressing either CD4 or CD8 ('single-positive' cells) develop here.

Thymic epithelial cells are a network of APCs expressing MHC class II that extend throughout the cortex and medulla and are involved in the selection of the T-cell repertoire.

Hassall's corpuscles are whorled structures of epithelial cells seen in the medulla of human thymus. Their function is uncertain.

T-CELL DEVELOPMENT

Education of T cells occurs in the thymus. Pre-T cells seed the thymus from the bone marrow and proliferate in the subcapsular region. These cells are $CD4^{-}8^{-}$ ('double-negative'), but they develop into the rapidly proliferating $CD4^{+}8^{+}$ cortical population, which constitutes the majority of thymocytes. They generate their antigen receptors (TCRs) and undergo positive and negative selection. The differentiating thymocytes lose either CD4 or CD8, leaving mature T cells expressing only CD4 or CD8 ('single-positive'), seen in the medulla. Cells that fail to generate a functional TCR, that cannot interact with self MHC, or that recognize self antigens die during development in the cortex and are phagocytosed by tingible body macrophages.

Positive and **Negative selection** are the processes by which thymocytes are rescued from apoptosis during development. Cells are positively selected by interaction with MHC molecules on thymic epithelial cells, and negatively selected if they recognize a self antigen presented by MHC molecules on dendritic cells acting as APCs.

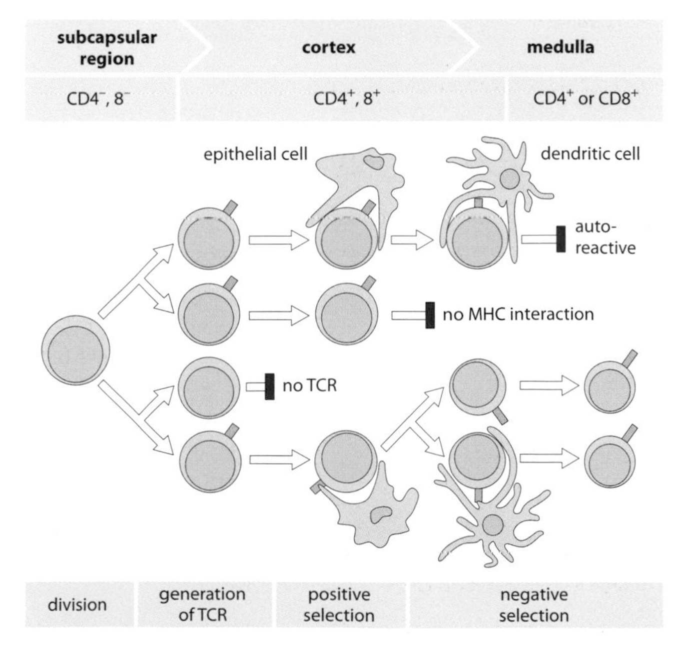

Fig. 1.13 T-cell development in the thymus.

LYMPH NODES

Lymph nodes are encapsulated organs that punctuate the lymphoid network and contain aggregations of lymphocytes and APCs. They are strategically placed to intercept antigens from the periphery; there are large groups of lymph nodes in the axillae, groin, and neck. The mesenteric lymph nodes are large and well suited to protect the body from antigens and pathogens from the gut. Lymph nodes are structurally organized into different areas:

Marginal sinus lies immediately beneath the capsule and is lined by phagocytic cells, the marginal zone macrophages, which can trap antigens entering the node.

Cortex, the outer region of the lymph node, contains mainly B cells. Follicles lie within this region.

Paracortex contains mainly T cells, interspersed with interdigitating dendritic cells expressing high levels of MHC class II molecules, that present antigen to T cells. The paracortex contains high endothelial venules (HEV), which are specialized vessels located in the lymphoid tissues. Large numbers of lymphocytes migrate across the HEV, which express specialized chemokines (such as CCL21) and adhesion molecules (such as GlyCAM-1).

Medulla contains relatively fewer lymphocytes and more macrophages and plasma cells than other regions. Medullary cords are strands of lymphocytes, which extend into the medulla.

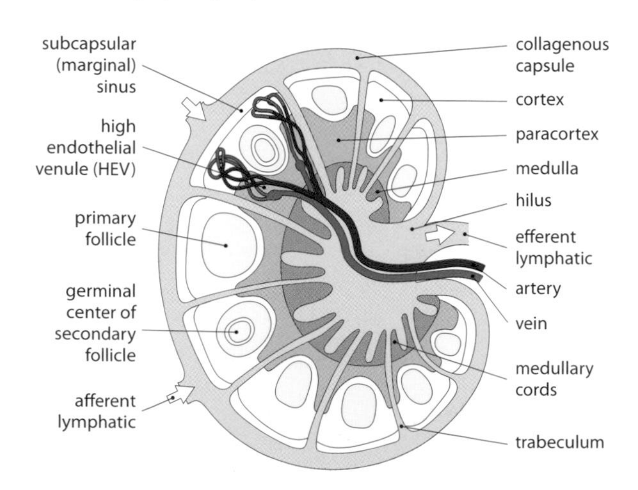

Fig. 1.14 The structure of a lymph node.

Afferent and Efferent lymphatics Cells arrive in the lymph nodes via HEV from the blood and afferent lymphatics, which drain into the marginal (subcapsular) sinus. They migrate into distinct areas and finally leave by the efferent lymphatic vessel.

Lymphoid follicles are aggregations of closely packed lymphocytes and APCs. Unstimulated lymph nodes contain primary follicles, which develop into expanded secondary follicles after antigen stimulation.

Germinal centers are regions of rapidly proliferating cells seen in the center of secondary follicles, which are important in the development of B-cell memory and the secondary antibody response. A few B cells initiate the germinal center and undergo rapid division in the basal dark zone (centroblasts). This is associated with somatic mutation of their immunoglobulin genes. The diversified cells become centrocytes in the basal light zone, where they may take up antigen released by follicular dendritic cells. B cells with high-affinity antibody are selected, whereas those with low-affinity antibody die and are phagocytosed by macrophages. Antigen-activated B cells move to the edge of the germinal center, to present antigen to $CD4^+$ T cells. They then undergo a second phase of division before leaving via the mantle zone, to become memory cells or plasma cells.

Bcl-2 is a molecule induced on centrocytes that have taken up antigen. Ligation of Bcl-2 rescues the cell from apoptosis. Bcl-2 is also expressed on developing hemopoietic cells in bone marrow.

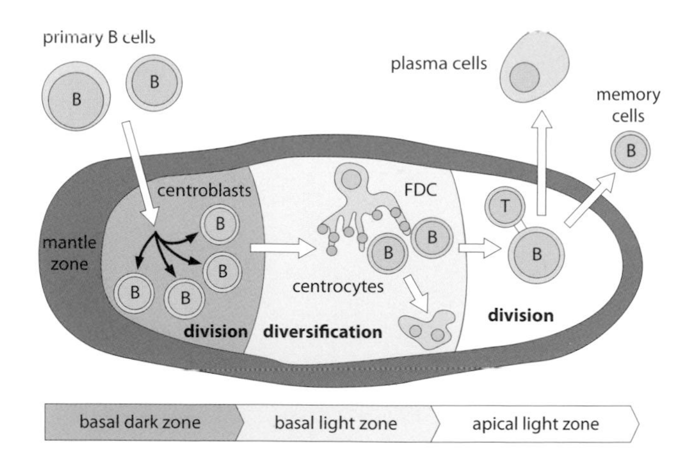

Fig. 1.15 B-cell development in a germinal center.

SPLEEN

The spleen is an encapsulated secondary lymphoid organ that lies in the peritoneum, beneath the diaphragm and behind the stomach. It contains two types of tissue, termed the red pulp and the white pulp or periarteriolar lymphatic sheath (PALS).

Red pulp consists of a network of splenic cords and venous sinuses lined by macrophages, which effect the destruction of effete erythrocytes. Plasma cells may also be seen in this region.

White pulp/PALS (Periarteriolar lymphatic sheath) contains the majority of the lymphoid tissue, distributed around the arterioles. T cells are found mainly around the central arterioles, and B cells further out. The B cells may be organized into primary and secondary lymphoid follicles with germinal centers. Phagocytes and antigen-presenting cells are also present in the follicles.

Marginal zone is the outer region of the PALS. It contains slowly recirculating B cells, CD169$^+$ marginal zone metallophils (MZMs), and macrophages, which present T-independent antigens to B cells. Marginal sinuses lie at the edge of the marginal zone. Most lymphocytes enter the PALS via specialized capillaries in the marginal zone and migrate out via bridging channels between the marginal sinuses into the venous sinuses of the red pulp.

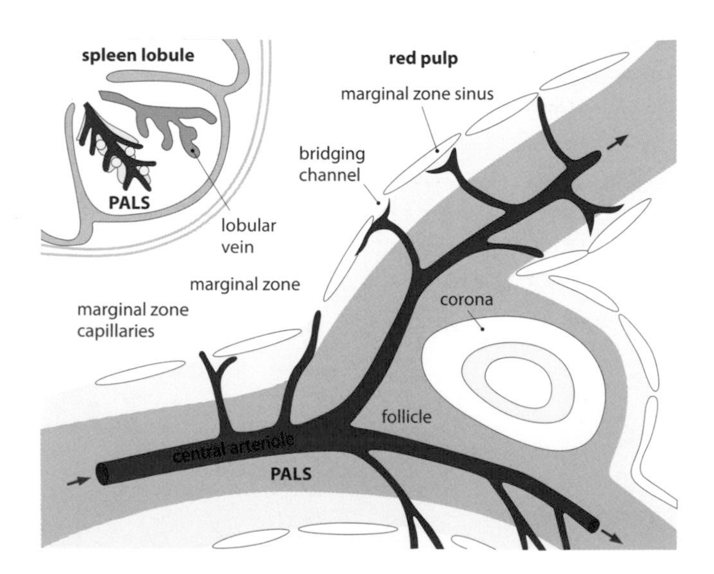

Fig. 1.16 The periarteriolar lymphatic sheath (PALS).

GUT-ASSOCIATED LYMPHOID TISSUES (GALT)

The GALT comprises the mucosa-associated lymphoid tissues of the gut. They include the focal accumulations of lymphocytes in the lamina propria and Peyer's patches, which contain high numbers of IgA-producing B cells and plasma cells. TREG cells secreting IL-10 and TGF-β are common in the lamina propria, helping to control immune responses against dietary antigens.

Peyer's patches are collections of lymphocytes in the wall of the small intestine, which appear as pale patches on the gut wall. The adjoining part of the intestinal mucosa lacks goblet cells and has a specialized epithelium that includes a unique cell type, the M cell, which transports antigens to the underlying lymphocytes. Cells enter a patch via the HEV, which selectively expresses an adhesion molecule MAdCAM-1 that binds lymphocytes expressing $\alpha_4\beta_7$ integrin. Lymphocytes exit from the patches via local lymphatics and selectively localize to the lamina propria.

Secretory immune system refers to the immune defenses present in secretory organs, such as salivary glands, lachrymal glands, mammary glands, and GALT. Their main protection is provided by secretory IgA. Dimeric IgA binds to a poly-Ig receptor on the basal surface of epithelial cells and is transported across the epithelium.

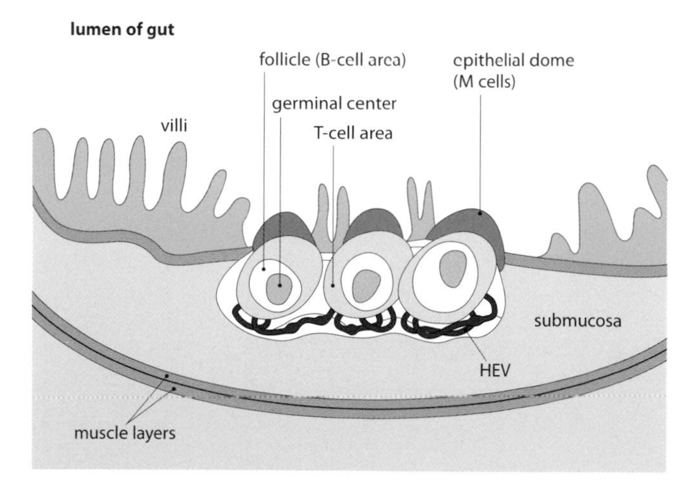

Fig. 1.17 Structure of a Peyer's patch.

2 Immune Recognition

ANTIGEN RECEPTORS

The immune system has two main ways of specifically recognizing antigen. B cells recognize intact antigens by using immunoglobulin (antibody) as their receptor. In contrast, T cells recognize antigen originating from within other cells, using their T-cell receptor (TCR). Many cells also have receptors for components of microbial pathogens that antedate the evolution of lymphocytes.

Antigen is the term used to describe any molecule that can be recognized by B cells or T cells. In general, immunoglobulins recognize and bind to intact antigens or large fragments that have retained some tertiary structure. Most T cells only recognize polypeptide fragments of antigens that have become associated with molecules encoded by the major histocompatibility complex (MHC) and that are expressed on the surface of other cells of the body.

Antigenic determinants, or epitopes, are the parts of an antigen to which an immunoglobulin binds. Antigens usually have many determinants, which may be different from each other or may be repeated molecular structures. Virtually the entire surface of a protein is potentially antigenic. Figure 2.1 illustrates epitopes on lysozyme recognized by three different monoclonal antibodies.

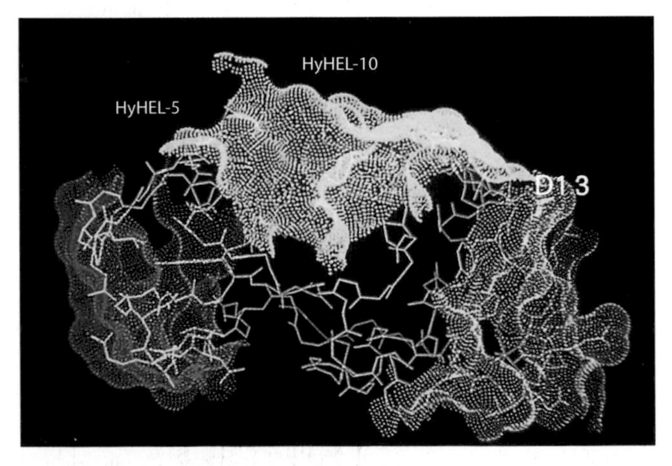

Fig. 2.1 Three epitopes of lysozyme. Courtesy of D. R. Davis.

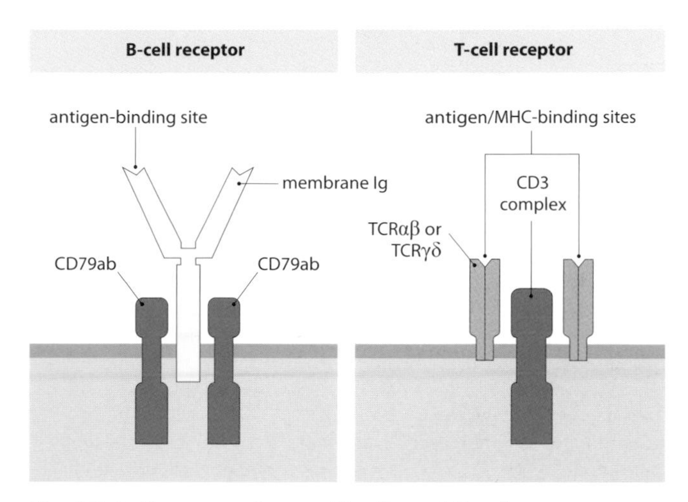

Fig. 2.2 Antigen receptors on B cells and T cells.

Antibodies (Ab)/Immunoglobulins (Ig) were originally identified as a class of serum proteins induced after contact with antigen, which bind specifically to the antigen that induced their formation. Subsequently it was found that B cells use a membrane-bound form of their secreted antibody as their antigen receptor (BCR). Surface immunoglobulins on B cells are associated with two polypeptides, Igα and Igβ (CD79a and CD79b).

Igα and Igβ (CD79) are transmembrane molecules that transduce activation signals to the B cell and are required for the expression of membrane Ig. Hence CD79 is a marker of mature B cells.

T-cell antigen receptors (TCRs) are integral membrane proteins on all mature T cells that specifically recognize antigenic peptides associated with MHC-encoded molecules. The receptor consists of a heterodimer responsible for MHC/antigen binding and a cluster of associated membrane-bound polypeptides, the CD3 complex, which triggers cell activation. The MHC/antigen-binding portion of the TCR varies between different clones of T cells, but the peptides of the CD3 complex (γ, δ, ε2, ζ2) are invariant.

Immunoreceptor tyrosine activation motifs (ITAMs) are segments found in the intracytoplasmic portion of many immune receptors, including CD79 and CD3, which are targets for phosphorylation by tyrosine kinases. ITAM phosphorylation promotes cell activation. Analogous inhibitory motifs (ITIMs) block cell activation if they become phosphorylated.

ANTIBODY STRUCTURE

Heavy chains and **Light chains** Antibody molecules all have a basic four polypeptide chain structure consisting of two identical light (L) chains and two identical heavy (H) chains, stabilized and cross-linked by intra- and inter-chain disulfide bonds (red) (Fig. 2.3). The heavy chains are glycosylated (between the C_H2 domains of IgG). There are five major types of Ig heavy chain (μ, γ, α, ε, δ), consisting of 450–600 amino acid residues, and the type determines the class of antibody. Light chains are of two main types (κ, λ), consisting of about 230 residues. Either type of light chain may associate with any of the heavy chains. Both heavy and light chains are folded into domains.

Pre-B cell receptor is a receptor found on developing B cells consisting of one μ heavy chain and surrogate light chains.

Membrane and Secreted immunoglobulins (Ig) Antibodies can be produced either as integral membrane proteins of mature B cells, which act as their antigen receptor, or in a secreted form. Secreted Igs are structurally identical to their membrane counterparts, except that they lack the transmembrane segment and short intracytoplasmic segment at the C-terminus. Secreted Igs are present in extracellular fluids and secretions. Virgin B cells produce membrane Ig, but after activation by antigen and differentiation into plasma cells they switch to the production of secreted immunoglobulins.

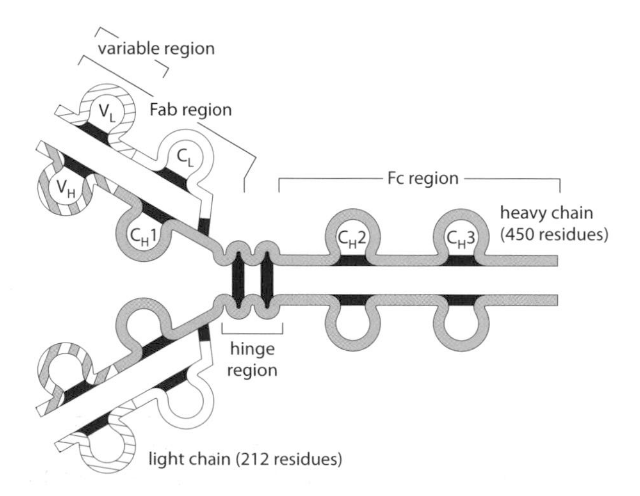

Fig. 2.3 Polypeptide chain structure of IgG1.

Variable (V) and **Constant (C) regions** Examination of the degree of amino acid variability between different antibody molecules of the same class shows that the largest amount of sequence variation is concentrated in the N-terminal domains of the light and heavy chains; this is therefore called the V region. The V regions of one light and one heavy chain form an antigen-binding site. The remaining domains are relatively invariant within any particular class of antibody, and so are called the constant (C) region. The domains of antibody molecules are named according to whether they are variable or constant and according to whether they are in the light or heavy chain. For example:

> **V_H** and **V_L** are variable domains of heavy and light chains.
>
> **C_L** and **C_H1** are the constant domains of the light chain and the first constant domain of the heavy chain, respectively.
>
> **C_γ, C_μ** etc. are domains of the heavy chain, which indicate the class of antibody. For example, $C_\mu1$ is the first constant domain of the μ heavy chain of IgM antibody.

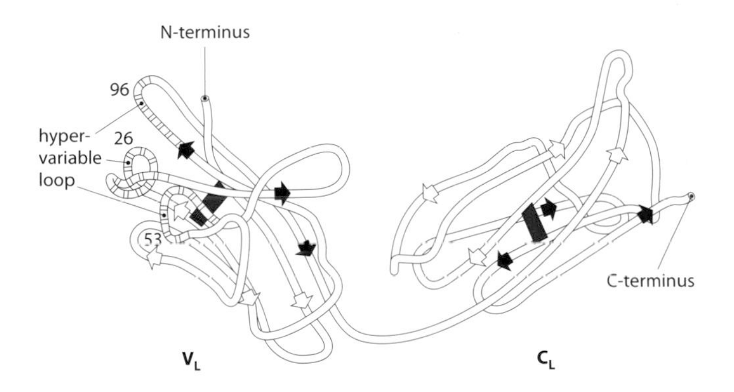

Fig. 2.4 The folding pattern of an immunoglobulin light chain.

Fab and **Fc regions** refer to two regions of the antibody, corresponding to antibody-binding (Fab) and constant (Fc) regions. The nomenclature originally refers to fragments produced by digestion with the enzyme papain.

Hinge region is a section of the heavy chain that contains inter-heavy chain disulfide bonds and confers segmental flexibility on the antibody molecule, so that both antigen-binding sites can independently engage the surface of a pathogen.

ANTIBODY–STRUCTURAL VARIATIONS

Classes and **Subclasses (Isotypes)** Antibodies may be grouped on the basis of structural similarities into different classes and subclasses, depending on their heavy chains. Each class serves different functions. In mammals there are five antibody classes: IgG, IgM, IgA, IgD, and IgE. IgG and IgA are further divided into subclasses. The number of subclasses varies between species. For example in humans there are four IgG subclasses, IgG1 to IgG4. As there is a gene in every individual for every one of the classes and subclasses, these are isotypic variants or antibody isotypes.

Kappa and **Lambda chains** Antibody light chains may also be divided into two types, namely κ and λ, which are encoded by separate gene loci. They, too, are isotypic variants. Either type of light chain can combine with one of the heavy chains.

Allelic exclusion is the process by which a cell uses either the gene from its maternal chromosome or the one from the paternal chromosome, but not both. Individual B cells display allelic exclusion of their heavy and light chain genes. T cells also display allelic exclusion of their TCR αβ or γδ heterodimers.

Single-chain antibodies Species of Camelidae (camels and llamas) produce antibodies consisting of paired heavy chains (no light chains). These antibodies demonstrate that a heavy chain alone can form an effective antigen-binding site and they have acted as models for the development of genetically engineered antibodies.

Single-domain antibodies (nanobodies) are single antibody V domains produced by genetic engineering, which can be used for applications that require immune recognition or targeting.

Idiotypes (Ids) are variants caused by the large amount of structural heterogeneity in the immunoglobulin V regions, which is related to the diversity required to bind different antigens. Some idiotypes are only made by animals that have a particular set of Ig genes (haplotypes) and these are 'germline idiotypes.'

Recurrent and **dominant idiotypes** Sometimes a particular idiotype is frequently seen in the immune response of different individuals to a particular antigen. This is a recurrent idiotype. If an idiotype constitutes a major part of an antibody response to that antigen, then it is a dominant idiotype.

Idiotopes are antigenic determinants on the V regions of antibodies that can be recognized by anti-idiotypic antibodies. An idiotype is identified by the collection of idiotopes it expresses. Individual idiotopes may be present on more than one antibody.

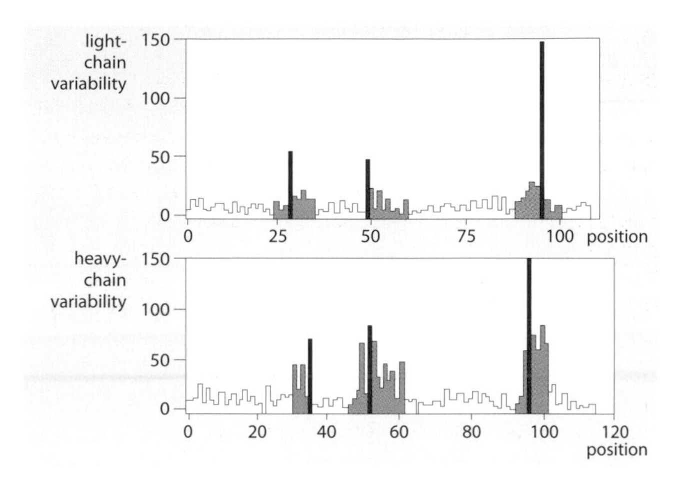

Fig. 2.5 Kabat and Wu plot of antibody heavy and light chains.

Kabat and Wu plot shows the amino acid sequence variability in immunoglobulin, determined by comparing the amino acid sequences of many different antibodies. It plots variability against amino acid position, thereby highlighting the most variable regions of the heavy and light chains.

Hypervariable regions and Framework segments. Within the V domains of the heavy and light chains there are three regions of greatest variability, which are clustered at the antigen-binding site. (red in the Kabat and Wu plot above). These hypervariable segments are separated by relatively invariant framework segments.

Complementarity-determining regions (CDRs) are the parts of the V domains that form the antigen-binding site. V-domain folding brings the CDRs together at the distal tip of the molecule.

Allotypes are variants due to intraspecies genetic differences. Each individual has a particular variant at each Ig gene locus, which will often differ from those in other individuals. In humans the Gm series of allotypes is found on IgG heavy chains.

Immunoglobulin supergene family (IgSF) The domain structure seen in antibodies consisting of three or four polypeptide loops stabilized by β-pleated sheet and a disulfide bond (sometimes called a β-barrel) is found in many molecules, which all belong to the immunoglobulin supergene family (see Fig. 2.4). The domain is seen in cell-surface receptors, including CD2, CD4, CD8, the T-cell receptor (TCR), MHC molecules, and Fc receptors (CD16, CD32, CD64).

ANTIBODY FUNCTIONS

Antibodies are bifunctional molecules. Their first function is to bind antigen and their second is to interact with host tissues and effector systems to facilitate removal of the antigen. Some antibody functions can be mediated just by binding to the antigen. For example, antibodies against surface molecules of viruses can prevent them from binding to and infecting host cells. However, most antibody functions require the antigen:antibody complex to bind to Fc receptors on cells. The antigen-binding sites are formed from the V domains of a heavy and light chain, whereas the C domains of the Fc region interact with cells of the immune system and C1q of the complement system. The different antibody classes and subclasses interact with different cells and so have different functions.

IgG is the major serum Ig and constitutes the main antibody in secondary immune responses to most antigens. In humans it is transferred across the placenta to provide protection in neonatal life. All IgG subclasses, except IgG4, can bind to C1q by sites in $C_\gamma2$ to activate the complement classical pathway. IgG can act as an opsonin by cross-linking immune complexes to Fc receptors on neutrophils and macrophages. It can also sensitize target cells for destruction by large granular lymphocytes with Fc receptors.

FcRn is the placental Fc receptor that binds all subclasses of IgG and transports them to the fetal circulation.

IgM is a pentamer of the basic four-chain structure. It is the first class to be produced during the development of the immune system and in the primary immune response. It fixes complement very efficiently and is the main antibody component of the response to T-independent antigens.

IgD is a trace antibody in serum but acts as a cell-surface receptor on many B cells, where it is co-expressed with IgM. IgD appears on differentiating B cells after activation, but is absent from mature antibody-forming cells.

IgA occurs as monomers, dimers, and polymers of the basic four-chain unit, existing in humans mostly as monomers and in other species as dimers. IgA is the most abundant Ig class in secretions, where it protects mucous membranes. It is also found in colostrum and is particularly important in protecting the neonates of species that do not transfer IgG across the placenta.

J chain is a polypeptide present in polymeric Igs (IgM and IgA), which facilitates their polymerization. It is synthesized by B cells but is not encoded by the Ig genes.

Poly-Ig receptor is present on the serosal surface of epithelial cells that transport and secrete IgA. It is a member of the Ig super-gene family, with five domains. IgA dimers bind to the receptor and are transported across the epithelium. The receptor is then cleaved, forming the secretory piece and releasing secreted IgA by exocytosis.

Secretory piece is the released form of the poly-Ig receptor, which attaches to IgA by disulfide bonds and is wound around the C domains of IgA to protect it from degradation by enzymes.

IgE binds to high-affinity Fc receptors (FcεRI) on mast cells and basophils, where it sensitizes them to release inflammatory mediators such as histamine after contact with antigen. IgE is particularly important in protection against helminthic infections, but it also mediates type I hypersensitivity reactions, such as asthma and hayfever.

immunoglobulin	heavy chain	mean serum concentration (mg/ml)	molecular weight (kDa)	number of heavy-chain domains	complement CI activation	placental transfer	epithelial transport	mast-cell binding
IgG1	γ1	9	146	4	+	+		
IgG2	γ2	3	146	4	+	+		
IgG3	γ3	1	170	4	+	+		
IgG4	γ4	0.5	146	4		+		
IgM	μ	1.5	970	5	+			
IgD	δ	0.03	184	4				
IgA1	α1	3.0	160	4				
IgA2	α2	0.5	160	4				
sIgA	α1 or α2	0.05	385	4			+	
IgE	ε	0.00005	188	5				+

Fig. 2.6 Properties of human immunoglobulin isotypes.

ANTIBODY GENES

The genes for antibodies lie at three gene loci on separate chromosomes; these are the *K*, *L* (κ, λ), and *H* (heavy-chain) loci. At each of these loci there are large numbers of different gene segments encoding polypeptides (exons), separated by segments that do not encode protein (introns) but contain sequences important in gene control and the process of recombination. The antibody genes undergo a number of recombinational events during B-cell development and maturation. The first events are DNA rearrangements of H and L chain genes, to form gene segments encoding their V domains.

Generation of diversity is the process by which large numbers of antibody V regions are generated. This is achieved by:

- Many different germline V genes in the *K*, *L,* and *H* loci
- Recombination between V, D, and J gene segments
- Insertion of nongermline (N) nucleotides into the joints
- Varied combinations of light and heavy chains
- Somatic mutation of V genes in individual B cells

T-cell receptors are diversified by similar mechanisms, although TCR genes are not subject to somatic mutation.

V genes encode the N-terminal 95 (approximately) amino acids of the antibody V domains. The number of V genes at each locus varies between loci and species. Analogous V genes are present in the four gene loci encoding TCR chains.

J genes and **D genes** To produce a gene encoding a heavy-chain V domain, any one of the H-chain V genes is recombined with any of a small number of D (Diversity) and J (Joining) genes to produce a VDJ gene. Recombination of light-chain genes is similar, except that they have no D-gene segments, and a V gene is recombined directly to a J gene. The T-cell receptor B and D loci have analogous D and J genes, while the TCR A and G loci have only J genes. (Note that J genes should not be confused with J chains.)

Recombination signal sequences (RSSs) and the **12/23 rule** Somatic recombination is the process by which the various gene segments for antigen receptors are brought together and joined. This process depends on specific recombination sequences flanking each V, D, and J gene that appose the segments, which are then enzymatically cut and rejoined to remove the intervening introns. The sequences consist of a heptamer, 12 or 23 bases, and a nonamer. The 12/23 rule states that a flanking sequence with 12 bases can only recombine with one of 23 bases. This ensures that heavy chains only make VDJ recombinations and light chains only make VJ recombinations.

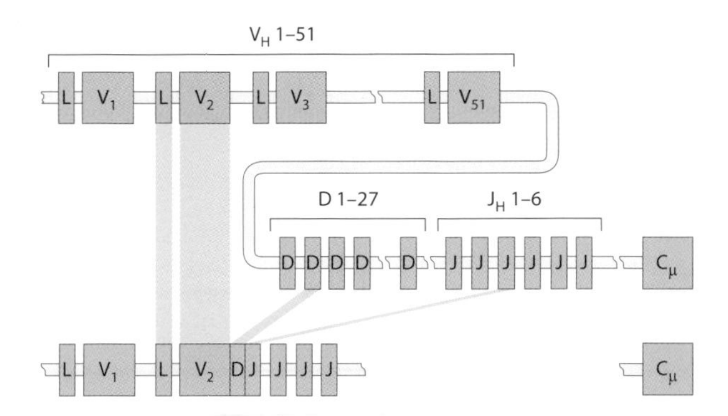

Fig. 2.7 VDJ recombination in the human IGH locus.

Junctional diversity is created when the ligation point in antibody gene recombination (VJ, VD, DJ) differs between B cells using the same gene segments. The shift produces different codons—see the sequence of the antibody S107 in Fig. 2.8.

N-region diversity is created when additional nucleotides are inserted into the gap between recombining gene segments with no corresponding germline DNA sequence—see the sequence of antibody M167 in Fig. 2.8. The reading frame must be restored to produce a functional antibody.

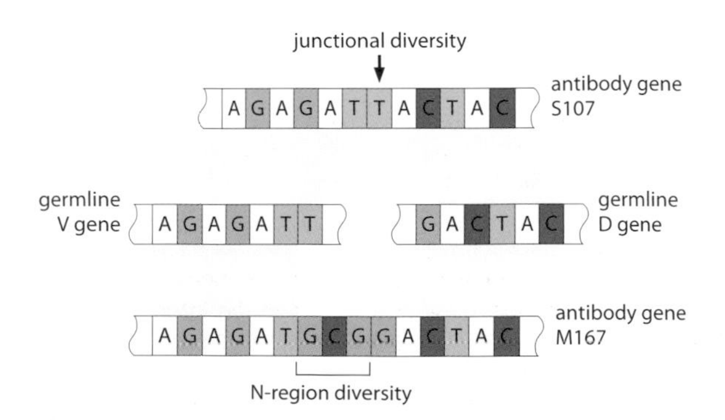

Fig. 2.8 Junctional and N-region diversity at a VD junction.

RAG-1, RAG2 (Recombination-activating genes) control the recombination of the TCR gene in T cells or the Ig genes in B cells. The enzymes recognize the recombination signal sequences and bring them together to initiate recombination by producing a double-strand break.

Terminal deoxynucleotidyl transferase (Tdt) is an enzyme that can add nucleotides to the exposed ends of the DNA during the recombination; these nucleotides become incorporated into the junctions between the V, D, and J gene segments.

Somatic hypermutation is the process by which DNA base changes occur during the lifetime of a B cell, producing point mutations in the Ig polypeptides. The high rate of mutation is centered on the recombined VJ and VDJ genes. The mechanism is activated in centroblasts and associated with class switching—IgG molecules usually vary more from germline sequences than IgM.

Antibody synthesis The segment of DNA encoding the recombined VDJ (heavy chain) or VJ (light chain) region and the C domains is transcribed into a primary RNA transcript that still contains introns. The transcript is then spliced to remove the introns, a process that involves recognition of sequences (donor and acceptor junctions) flanking the exons. This leaves mRNA, which is translated across the membrane of the endoplasmic reticulum (ER). Each mRNA has a leader or signal sequence, by which it is directed to the ER. The process is illustrated below for a membrane IgM μ polypeptide. Complete Igs are assembled and glycosylated in the ER and stored in the Golgi apparatus. Secreted Ig is released by exocytosis, whereas membrane Ig, associated with CD79-signaling peptides, is moved to the cell surface.

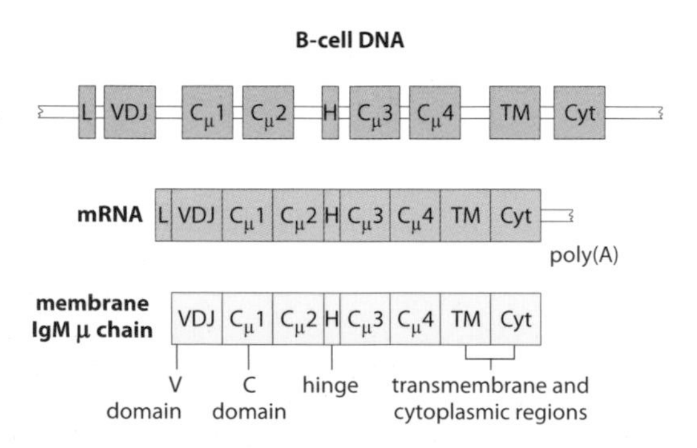

Fig. 2.9 Production of an IgM μ polypeptide.

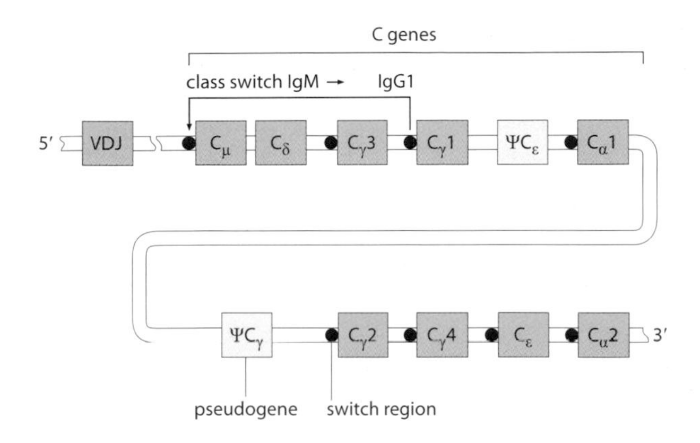

Fig. 2.10 Class switching in the human IGH C-gene locus.

C genes The heavy-chain constant-region genes are arranged downstream (3′) of the recombined VDJ gene. Each gene consists of a series of exons encoding the individual C domains, as well as separate exons for the hinge (except IgA) and for the transmembrane and cytoplasmic regions. The primary transcript of the heavy chains can be processed, to produce mRNA either for membrane or for secreted Ig. To produce membrane Ig, the exons for the transmembrane segments are spliced to a point just within the final C domain. If this does not occur, the stop signal is retained and mRNA for secreted Ig is produced. The point of polyadenylation controls how the primary transcript will be spliced. Initially a B cell joins a μ gene to its VDJ gene, but other C genes may displace the μ gene in a process called class switching.

Class switching is a process by which a B cell can switch the class of Ig it produces while retaining the same antigen specificity. All the heavy-chain constant-region genes except Cδ are preceded by a switching sequence. Switching is effected by bringing a new C gene up to the position occupied by the μ C gene, with the loss of the intervening C genes. This process is illustrated above for the switch from IgM to IgG1. It is also possible for a cell to switch classes by producing very long primary transcripts, which are then spliced to connect the new C gene to VDJ. Indeed, this is the only way in which IgD (which lacks a switch sequence) can be produced. The process is controlled by T cells and modulated by cytokines. For example, in humans IL-4 promotes switching to IgG4 and IgE, whereas IL-5 promotes a switch to IgA.

ANTIBODY BIOTECHNOLOGY

Much of the early work on the elucidation of antibody structure was performed using fragments of antibodies prepared by enzyme digestion. For example, the Fab and $F(ab')_2$ fragments of IgG are produced by digestion with papain or pepsin, respectively. Lacking the Fc region, they are both useful in determining which antibody functions are Fc-dependent. The development of monoclonal antibody technology was a major step forward in allowing researchers to produce large quantities of well-defined antibodies. More recently, genetic engineering has been used to generate antibodies and antibody fragments for specific applications.

Polyclonal and **monoclonal antibodies** Immunization of an animal stimulates antibody production from a large number of different clones of B cells. These antibodies will differ in their epitope specificity and affinity for the antigen. Such antibodies are referred to as 'polyclonal.' In contrast, the antibodies produced by a single clone of B cells (monoclonal antibodies) have a defined specificity and affinity. Note that a monoclonal antibody is not necessarily of higher affinity than a polyclonal antibody, and whether it is more effective than polyclonal antibody depends on the assay or purpose it is used for.

Phage display antibodies is a method for producing antibody fragments. Mixed mRNAs for antibody V_H and V_L domains are cross-linked with a spacer to give a gene for an Fv fragment. The gene is inserted into a vector (phage), which expresses the Fv on its tips. Phages are selected according to their binding specificity and transfected into bacteria to synthesize the Fv fragments.

Humanized antibodies are required where the antibody itself must not be antigenic, for example, for long-term therapy of patients. The genes for the antigen-binding hypervariable regions of the required antibody are spliced into the genes encoding the framework regions of a human heavy- or light-chain V-domain.

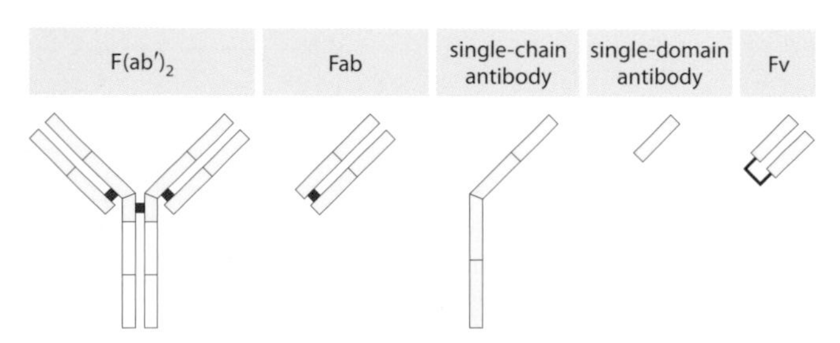

Fig. 2.11 Antibody fragments.

ANTIGENS

Immunogens An antigen is any molecule recognized by the immune system, but the term immunogen is reserved for those antigens that elicit a strong immune response, particularly in the context of protective immunity to pathogenic organisms.

Haptens and **Carriers** Artificial antigens have been used to examine the immune response. In particular, small antigenic determinants (haptens) are covalently coupled to larger molecules (carriers). Haptens bind to antibodies but cannot by themselves elicit an antibody response. Haptens are recognized by B cells, which present fragments of the carriers to T cells.

T-dependent antigens need to be recognized by both T cells and B cells to elicit an antibody response. Most protein antigens fall into this category. T-dependent antigens induce class switching to IgG and IgA with an increase in antibody affinity.

T-independent antigens can stimulate B cells to produce antibody without T-cell help. Most such antigens are large polymeric molecules, with repeated epitopes, capable of cross-linking surface Ig on B cells, and they are only slowly degraded.

Type I and **Type II T-independent** antigens are differentiated according to their ability to activate different B-cell subsets. Type I antigens stimulate both $Lyb5^+$ and $Lyb5^-$ cells in mouse, whereas type II antigens can only act on $Lyb5^+$ cells.

antigen	polymer	B-cell mitogen	resistance to degradation	type
lipopolysaccharide (LPS)	+	+++	+	1
PPD	–	+++	+	1
dextran	++	–	++	2
levan	++	–	++	2
Ficoll	+++	–	+++	2
polymerized flagellin	++	+	+	2
poly(I): poly(C)	++	++	+	2
poly-D-amino acids	+++	–	+++	2

PPD = Purified protein derivative of M. tuberculosis

Fig. 2.12 Properties of commonly used T-independent antigens.

ANTIGEN–ANTIBODY INTERACTIONS

Epitopes and **Paratopes** are part of a nomenclature used to describe the interaction between antigen and antigen-receptor molecules, including antibodies. An epitope is an antigenic determinant; a paratope, formed by hypervariable loops of V domains, is the part of the antibody that binds to the epitope.

Contact residues are the amino acids of the epitope and paratope that contribute to the antigen–antibody bond.

Continuous and **Discontinuous epitopes** Study of the molecular interaction between antigen and antibody shows that some of the epitopes are formed by one linear stretch of amino acids (continuous epitope). In most cases, however, an epitope has contact residues from different sections of an antigen brought together by folding of the polypeptide (discontinuous epitope).

Antigen–antibody bond Antibodies bind specifically to the antigen that induced their formation by multiple noncovalent bonds, including van der Waals forces, salt bridges, hydrogen bonds, and hydrophobic interactions. Crystallographic studies of immune complexes between antibodies and protein antigens indicate that they interact by complementary surfaces of up to 1000 $Å^2$ with the third hypervariable region (VJ, VDJ) lying near the centre of the binding site. Hypervariable regions of both L and H chains contribute contact residues. Figure 2.13 (top) shows lysozyme antigen (green) and the light (yellow) and heavy (blue) chains of complexed anti-lysozyme Fab. The lower diagram shows the molecules rotated forward through 90°, with contact residues (red) numbered on the interacting faces.

Charge neutralization refers to the observation that charged contact residues on an epitope are often neutralized by residues of an opposite charge on the paratope. This is particularly important at the center of the binding site.

Induced fit refers to the flexing of residues in the hypervariable loops in contact with the epitope, which may occur to allow optimum fit between the interacting molecules.

Antibody affinity is a measure of the bond strength between a single epitope and a paratope. It depends on the sum of the bond energies of the noncovalent interactions, set against the natural repulsion between molecules and the energy required to make any necessary distortions to allow binding (induced fit).

Antibody valency describes the number of binding sites on a molecule. For example, IgG has two sites and IgM has ten, although the actual number of bonds that can be formed depends on the configuration of the antigen.

Antibody avidity is the total strength of an antigen–antibody bond, which is related to the affinity of the paratope–epitope bonds and antibody valency. Binding energy is much enhanced when several bonds form, so avidity usually exceeds affinity.

Cross-reaction Some antisera are not totally specific for their inducing antigen, but bind related (cross-reacting) antigens, either because they share a common epitope or because the molecular shapes of the cross-reacting antigens are similar.

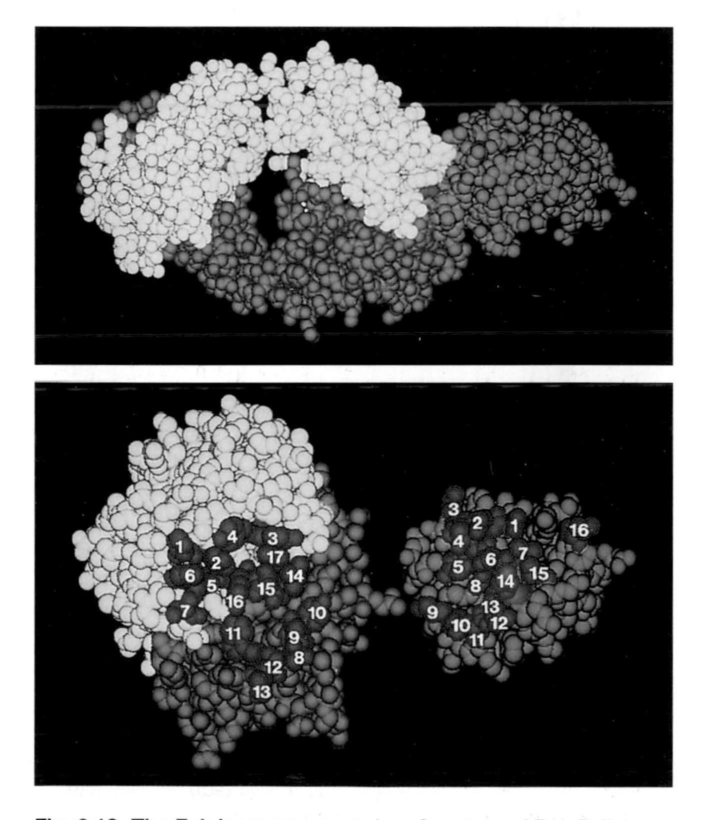

Fig. 2.13 The Fab:lysozyme complex. Courtesy of R.J. Poljak from *Science* 1986, 233:747. Copyright 1986 by the AAAS.

T-CELL ANTIGEN RECEPTOR (TCR)

The TH cell antigen receptor consists of a heterodimer (Ti) and a number of associated polypeptides that form the CD3 complex. The dimer recognizes processed antigen associated with an MHC molecule. The CD3 complex is required for receptor expression and is involved in signal transduction.

TCRαβ (TCR2) and **TCRγδ (TCR1)** The polypeptide chains for the antigen-binding portion of the receptor are encoded by four different gene loci: *TCRA*, *B*, *G*, and *D*. Any T cell will express either an αβ or a γδ receptor. The great majority of thymocytes and peripheral T cells have a TCRαβ.

Ti is a term used to distinguish the antigen:MHC-binding portion (which differs between cells) from the monomorphic CD3 complex. The N-terminal domains of αβ or γδ resemble a membrane-bound Fab, with variable (V) domains forming the antigen:MHC receptor and membrane-proximal constant (C) domains.

CD3 complex in humans consists of four polypeptide chains, each of which spans the cell membrane. These are the γ, δ, ε, and ζ chains. The first three are structurally related single-domain members of the Ig supergene family; the ζ chains are unrelated and form ζ–ζ dimers. In mice, a fifth chain, η, is also present as a minority alternative partner for ζ chains, making a η–ζ dimer. The CD3 ζ–ζ dimer has intracellular ITAM motifs, which become phosphorylated after the receptor binds to antigen:MHC, allowing it to bind to kinases, which initiate T-cell activation.

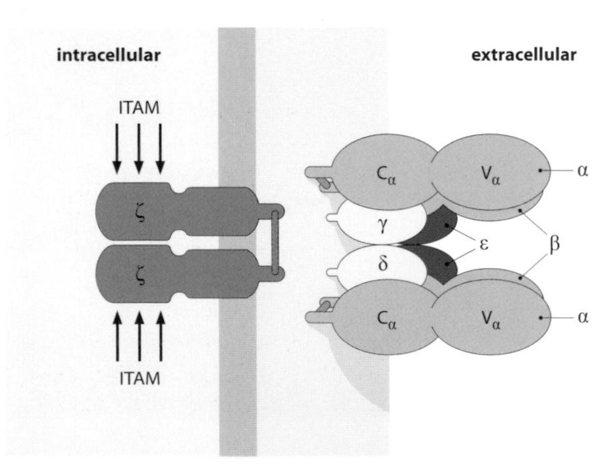

Fig. 2.14 A model of a T-cell receptor complex (TCR2).

T-CELL RECEPTOR GENES

The genes for the antigen:MHC-binding portion of the TCR are similar to those of antibody, in that they consist of multiple V, D, and J segments which become recombined during T-cell development to produce function VDJ or VJ genes (see p. 34). These encode the N-terminal V domains of the TCR. The α and γ loci have V and J segments only, whereas β and δ have V, D, and J segments. The recombined V gene is linked to the exons for the C domains, the short hingelike section (containing the interchain disulfide bond), the transmembrane and cytoplasmic segments. The layouts of the human α and β loci are shown below, and those of the mouse α, β, and δ loci are very similar. Note that there are tandem sets of genes for the β-chain D, J, and C regions. Each locus is distinct, although the δ-chain D, J, and C genes lie between the V_α and J_α genes. The process of recombination can permit variability in the precise linking position of V to J, the possibility of linking D segments in all three reading frames and the addition of N-region diversity—that is, insertion of bases not encoded in the germline. Theoretically, the arrangement of recombination sequences flanking the D_β and D_δ genes permits the assembly of genes with more than one D region (that is, VDDJ). In contrast to antibody genes, the TCR genes do not undergo somatic hypermutation. Nevertheless the amount of diversity that can be generated is at least as great as for antibodies. The genes for the γ, δ, and ε polypeptides of the CD3 complex do not rearrange and are closely linked on chromosome 11 in humans. All CD3 genes are required for TCR expression, and charged residues in the CD3 transmembrane segments are thought to be involved in association with the antigen-binding $\alpha\beta$ or $\gamma\delta$ dimers.

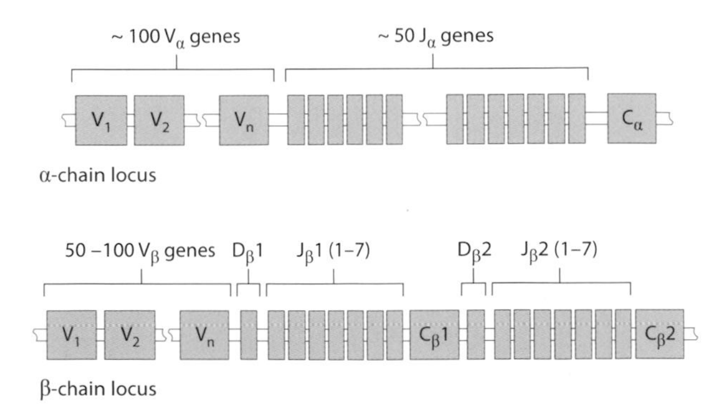

Fig. 2.15 Genes of the human *TCRA* (α) and *TCRB* (β) loci.

MHC MOLECULES

Major Histocompatibility Complex (MHC) is a large group of genes, including those encoding the class I and II MHC molecules, involved in the presentation of antigen to T cells. The complex was originally identified as a locus encoding allogeneic cell-surface molecules involved in graft rejection. A variety of other proteins are also encoded in the MHC, including complement components (C4, C2, FB), heat shock proteins, and cytokines (TNF-α, TNF-β).

MHC class I molecules are integral membrane proteins found on all nucleated cells and platelets. They are the classical transplantation antigens, each having one polypeptide chain encoded within the MHC that traverses the plasma membrane. The extracellular portion has three domains (α_1–α_3). The membrane-proximal α_3 domain is associated with β_2-microglobulin, whereas the two N-terminal domains form an antigen-binding pocket, consisting of a base of β-pleated sheet derived from both α_1 and α_2 domains, surrounded by two loops of α helix. Residues facing into the binding pocket vary between different molecules and haplotypes, to allow different antigenic peptides to bind. The α_3 domain has a binding site for CD8.

β_2-Microglobulin (β_2m) is a polypeptide encoded by a gene outside the MHC, which forms a single domain related to Ig domains. It is necessary for loading and transport of class I to the cell surface.

Class I-like (nonclassical, Ib) MHC molecules have the same basic structure as MHC class I molecules, and a variety of functions. Some are encoded within the MHC, but many are not.

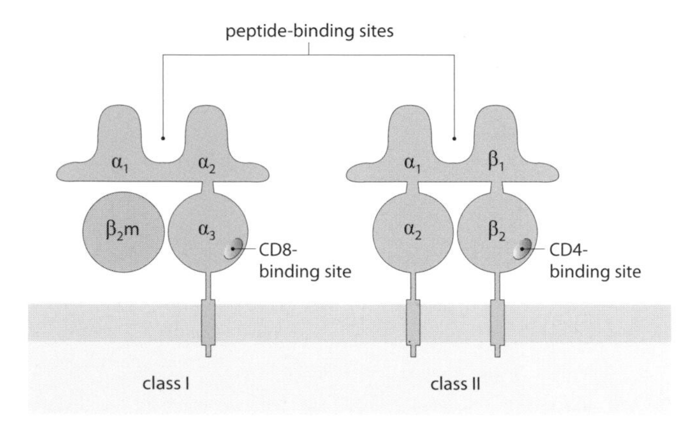

Fig. 2.16 Structures of MHC class I and class II molecules.

CD1 is a group of four MHC class I-like molecules with deep antigen-binding pockets that can accommodate acyl groups of glycolipid and lipoprotein antigens which they present to T cells, such as lipoarabinomannan from mycobacteria.

MHC class II molecules (Ia antigens) are expressed on B cells, macrophages, monocytes, APCs, and some T cells. They consist of two noncovalently linked polypeptides (α and β), both encoded within the MHC, which both traverse the plasma membrane, each having two extracellular domains. Class II molecules resemble class I molecules with the N-terminal α_1 and β_1 domains forming the peptide-binding site. Another site in the β_2 domain binds to CD4. Several class II-like genes (DM) are also encoded in the MHC. They facilitate the loading of antigenic peptides onto the class II molecules.

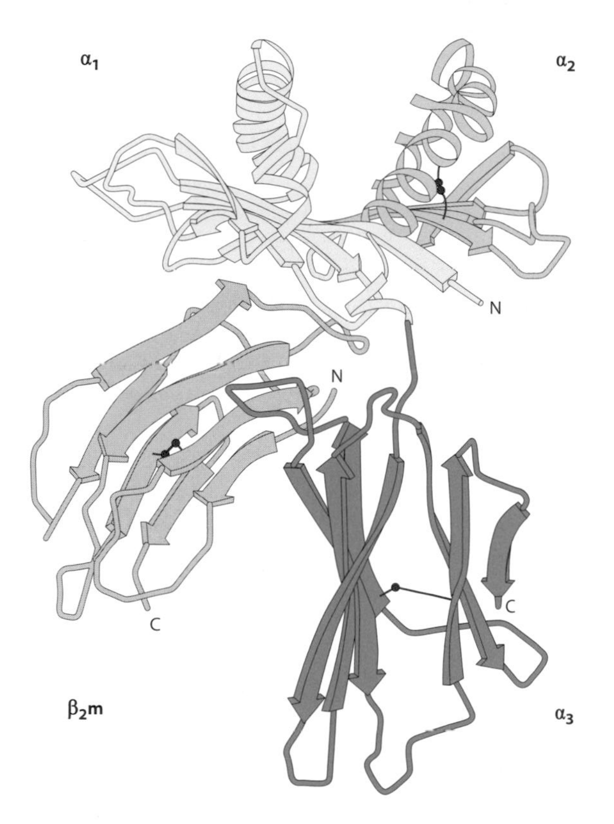

Fig. 2.17 Structure of a class I MHC molecule.

MHC GENES

A major histocompatibility complex is found in all mammal species. In humans the locus is called HLA; in mice it is the H-2 complex and in rats it is RT-1.

HLA (Human Leukocyte Antigen) locus is the human major histocompatibility complex, so called because the MHC molecules were originally identified as antigens on the surface of leukocytes and genetic variability in the MHC molecules was identified serologically. Nowadays variations are identified by genotyping. The HLA complex contains more than 220 individual gene loci, of which 21 have an immunological function. The class I and class II genes are highly polymorphic, with more than 6000 class I sequence variants and 1500 class II variants identified. There is also some variation in copy number in individual loci between haplotypes. The gene complex is located on chromosome 6, and it includes three principal class I and three class II loci.

HLA-A, -B and **–C loci** encode the α chains of the classical MHC class I molecules, expressed by all nucleated cells, which present antigens to $CD8^+$ cytotoxic T cells.

HLA-E encodes a class I-like molecule that presents the signal sequence (leader) peptides of the classical MHC class I molecules to NK cells. The complex is recognized by a receptor consisting of CD94 and NKG2. HLA-E genes have limited polymorphism.

HLA-G is a class I-like molecule expressed on the placental syncytiotrophoblasts (which do not express HLA-A, -B, and –C) and is thought to prevent allograft rejection of the fetus mediated by NK cells. It can be produced in membrane-bound and soluble forms.

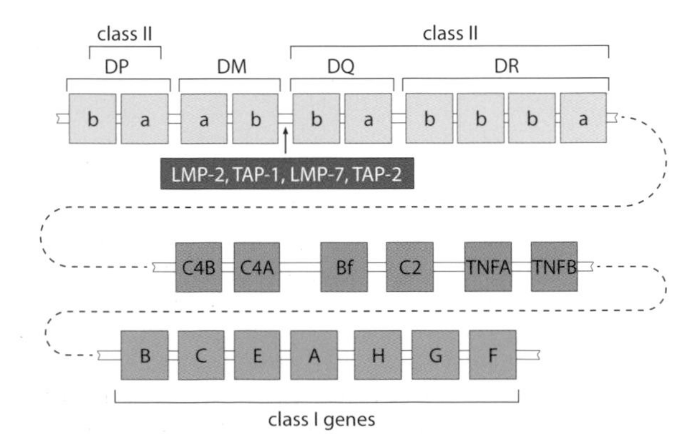

Fig. 2.18 HLA—the human major histocompatibility complex.

HLA-DP, -DQ, and **–DR** loci encode class II MHC molecules expressed on APCs, which present peptides to $CD4^+$ T cells. Originally these were described as HLA-D specificities, detected by their ability to stimulate allogeneic cells in mixed lymphocyte cultures. Later they were defined serologically and most recently by gene sequence. DP and DQ each encode one pair of class II α and β chains, plus pseudogenes. The DR locus encodes one non-polymorphic α chain and one to four β chains depending on the individual haplotype. Since α chains encoded on one chromosome can combine with β chains encoded on the other, this is a source of additional structural diversity in class II molecules.

HLA-DM encodes the class II molecule DM, which is involved in loading peptides onto class II molecules.

LMP-2 and **LMP-7** encode components of proteasomes which are induced by interferon-γ and modify the proteasome function.

TAP-1 and **TAP-2** encode transporters that take antigenic peptides from the cytoplasm into the endoplasmic reticulum.

HLA-class III genes is a catch-all term for other genes encoded within the MHC, including complement components C2 and FB, the pseudoalleles for C4 (C4F and C4S), which determine the Rogers and Chido blood groups, respectively. Genes for TNF, some heat shock proteins (such as HSP7), and enzymes (such as adrenal steroid 21-hydroxylase, CYP21) lie in this region.

H-2 is the mouse major histocompatibility complex, which lies on chromosome 17. There are six main regions: K,M,A,E, S, and D.

H-2K and **H-2D** encode class I MHC molecules. The K locus has one gene, whereas the number of genes in the D locus varies between strains.

H-2A and **H-2E** encode the α and β chains of the class II molecules. This was previously designated as the H-2I region and subdivided into I-A and I-E.

H-2S includes the genes for complement components and is analogous to the 'class III' region in humans.

H-2T region (Qa and **Tla loci)** lies downstream of the main H-2 complex and contains genes for more than 25 class-I like molecules. Some function as hemopoietic differentiation molecules; others present antigens or interact with NK cells. Some of them may be pseudogenes that act as a source of DNA for gene conversion with conventional class I molecules, to promote gene diversity. Some of the genes were originally identified on thymocytes or as thymic leukemia antigens (Tla).

INNATE IMMUNE RECOGNITION

Pathogen-associated molecular patterns (PAMPs) are common molecular motifs found on a number of pathogens. Examples are bacterial flagellin and double-stranded RNA. These motifs can be used to recognize pathogen infection.

Pattern recognition receptors (PRRs) is the generic term for cell-surface receptors and soluble molecules that recognize PAMPs. Many of these receptors are evolutionarily ancient (such as the Toll-like receptors TLR), and they are expressed on many different cell types. Mononuclear phagocytes have a particularly wide range of pattern recognition receptors.

Mannose receptor (CD206) is present on macrophages, monocytes and a subset of dendritic cells. The receptor contains eight C-type lectin domains that can bind carbohydrate groups containing mannosyl or fucosyl residues, and a terminal lectin domain that binds sulfated carbohydrate groups (Fig. 2.19). The receptor can recognize a number of microbial proteoglycans, but also binds endogenous proteins, including myeloperoxidase, lysosomal hydrolases, and some hormones.

Scavenger receptors are a structurally diverse group of receptors present on macrophages, dendritic cells, and some endothelial cells. Three receptors belonging to the SR-A family (SR-AI (CD204), SR-AII, and MARCO), bind to components of Gram-positive and Gram-negative bacteria, including lipopolysaccharide and lipoteichoic acids. They contribute to the ability of macrophages to phagocytose bacteria and promote clearance of apoptotic cells.

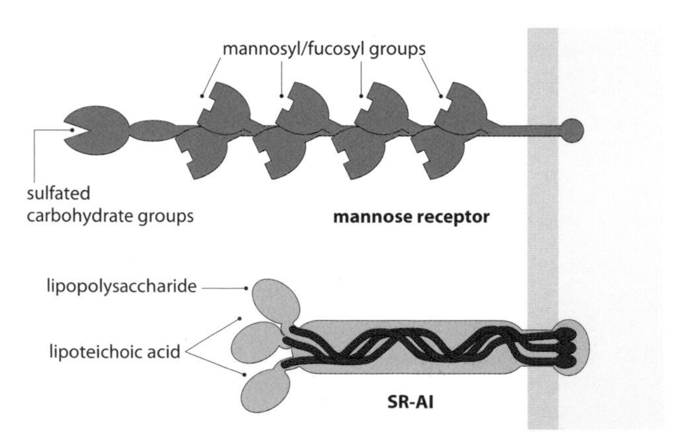

Fig. 2.19 Mannose receptor and scavenger receptor SR-AI.

Siglecs are a family of 12 proteins that bind sialic acid (sialic acid-binding, Ig-like lectins). Siglec-1 (sialoadhesin, CD169) is strongly expressed on macrophages in lymphoid tissues and less strongly on other tissue macrophages. It is thought to mediate intercellular adhesion by binding to extracellular matrix and other cell-surface molecules, including leukosialin (CD43) and the mannose receptor. Because sialic acid is expressed on eukaryotic cells but not on most microbes, it can distinguish them, and some of the siglecs inhibit immune activation. Siglec-2 (CD22) expressed on B cells is associated with the receptor complex (BCR) and promotes endocytosis. Siglec-3 (CD33) expressed on macrophages and myeloid stem cells also belongs to this family.

Dectins are receptors on macrophages and dendritic cells with a single lectin-like domain. Dectin-1 binds β-glucan from fungi and promotes their phagocytosis. Individuals lacking Dectin-1 are susceptible to mucosal candidiasis.

DC-SIGN (Dendritic cell-specific ICAM3-grabbing non-integrin) is a mannose-binding C-type lectin found on dendritic cells and some macrophages. It interacts with Toll-like receptors and is thought to promote signaling between APCs and T cells.

MINCLE (Macrophage-inducible C-type lectin) recognizes fungal pathogens, as well as components of necrotic cells. It signals via the ITAM-containing γ chain of Fc receptors (FcRγ).

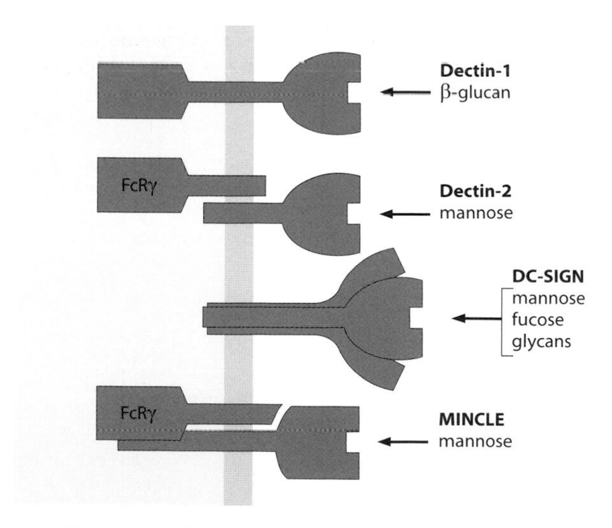

Fig. 2.20 Lectin-like receptors.

Toll-like receptors (TLRs) are a family of receptors involved in the recognition of a wide range of microbial molecules (Fig. 2.21). The prototypic receptor Toll was first identified in the fruitfly *Drosophila*, but several TLRs are found in mammals, particularly on mononuclear phagocytes. Each receptor recognizes a small range of conserved molecules from a group of pathogens. Most of them are located at the cell surface, but TLR3, 7, 8, and 9, which recognize viral components, are on endosomes. The TLRs have an intracellular domain, similar to that on the IL-1 receptor. Ligation of TLRs activates cells, leading to the production of inflammatory cytokines, including TNF-α and IL-12. It also enhances the cells' antimicrobial killing mechanisms and antigen-presenting capacity. Signals from TLRs potentiate macrophage activation by IFN-γ.

TLR2 can form heterodimers with TLR1 or TLR6, generating receptors that recognize a variety of microbial components.

TLR4 is the best-characterized of this family of receptors. It binds to LPS as well as a number of host protein molecules that are released at sites of damage or infection, such as heat shock protein-60 (HSP60), and a variant of fibronectin produced at sites of inflammation.

CD14 and LPS-binding protein The binding of LPS to TLR4 depends on two additional proteins: CD14, a cell-surface molecule of macrophages that acts as a co-receptor for LPS, and LPS-binding protein, a serum molecule that captures LPS and transfers it to CD14 (Fig. 2.22).

Receptor	Ligand	Pathogens recognized
TLR1	lipopeptides	Gram-negative bacteria, mycobacteria
TLR1/2	tri-acyl lipoprotein	bacteria
TLR2	lipoteichoic acid lipoarabinomannan zymosan GPI-linked peptides	Gram-positive bacteria mycobacteria fungi *Trypanosoma cruzi*
TLR2/6	di-acyl lipoprotein	bacteria
TLR3	dsRNA	viruses
TLR4	lipopolysaccharide	Gram-negative bacteria
TLR5	flagellin	bacteria
TLR6	di-acyl lipopeptides	mycobacteria
TLR7	ssRNA	viruses
TLR8	ssRNA	viruses
TLR9	unmethylated CpG	bacteria

Fig. 2.21 Properties of the Toll-like receptors (TLRs).

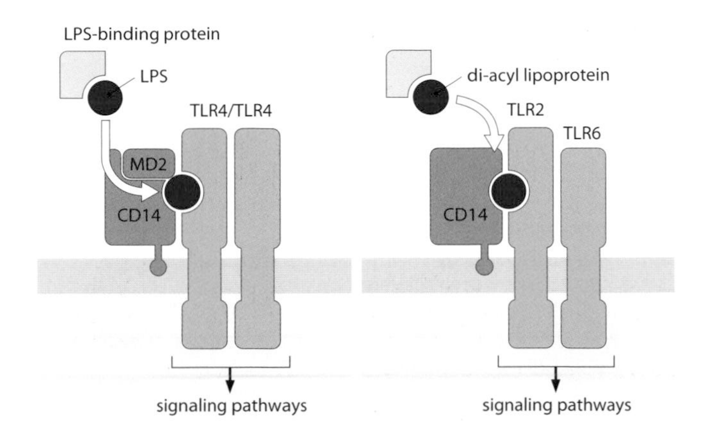

Fig. 2.22 Activation of macrophages by LPS.

Pentraxins are a group of soluble pentameric molecules that exhibit calcium-dependent binding to carbohydrates. The group includes C-reactive protein (CRP), serum amyloid-P (SAP) and pentraxin-3 (PTX3). Both CRP and SAP are primarily produced and broken down by the liver.

C-reactive protein (CRP) is an acute-phase protein that increases rapidly in the serum during inflammation and is used as a clinical marker of inflammation. It binds to phosphocholine groups on pneumococci, which opsonizes them and promotes their phagocytosis by macrophages, both directly and by activating complement.

Serum amyloid-P (SAP) recognizes a number of products of tissue breakdown, including amyloid fibers.

Ficolins are a group of three soluble lectins. Ficolin-1 (FCN1), secreted by mononuclear phagocytes, recognizes components of the cell wall of Gram-positive bacteria and activates the lectin pathway of complement, to opsonize them. Ficolin-2 (FCN2) also recognizes components of the bacterial cell wall and apoptotic cells.

Collectins is the name for the complement components mannan-binding lectin and conglutinin, soluble pattern recognition receptors that can activate complement.

3 Immune Responses

ADAPTIVE AND INNATE IMMUNITY

The immune response is mediated by a variety of cells and soluble factors, broadly classified according to whether they mediate adaptive (acquired) or innate (natural) immunity.

Adaptive (acquired) immunity is specific for the inducing agent and is marked by an enhanced response on repeated encounters with that agent. Thus the key features of the adaptive immune response are memory and specificity.

Innate (natural) immunity depends on a variety of immunological effector mechanisms, which are neither specific for particular infectious agents nor improved by repeated encounters with the same agent. In practice, there is considerable overlap between these two types of immunity: antibodies can direct or activate elements of the innate system, such as phagocytes and complement. Receptors of the innate immune system, including phagocyte receptors, are described on pages 48–51. Other elements of innate immunity are outlined below.

Complement system is a group of serum molecules involved in the control of inflammation, the removal of immune complexes and lysis of pathogens or cells sensitized by antibody, or mediators of the collectin, ficolin, and pentraxin families.

Acute-phase proteins are serum molecules that increase rapidly at the onset of infection, such as C-reactive protein, serum amyloid-P, serum amyloid-A, and mannan-binding lectin (MBL).

	innate immune response	**adaptive immune response**
	resistance not improved by repeated infection	resistance improved by repeated infection
soluble factors	lysozyme, complement, acute-phase proteins such as CRP, interferons	antibody
cells	phagocytes natural killer (NK) cells	T lymphocytes

Fig. 3.1 Elements of the innate and adaptive immune systems.

Interferons (IFNs) are a group of molecules that limit the spread of viral infections. There are three types: IFN-α and IFN-β, produced by leukocytes and fibroblasts, and IFN-γ, produced by activated T cells and NK cells. Interferons from activated or virally infected cells bind to receptors on nearby cells, inducing them to make antiviral proteins. IFN-α and IFN-β bind to a type I IFN receptor, whereas IFN-γ binds to a type II receptor. IFN-γ also has many other immunomodulatory functions (see p.69).

Antiviral proteins are molecules that are induced by IFN, which limit viral replication. Many of them are produced in an inactive form and are activated in infected cells by contact with viral products such as dsRNA. Activated antiviral proteins include some that block the initiation of protein synthesis and others that cause mRNA degradation, thus reducing viral protein synthesis.

Cell-mediated immunity and **Humoral immunity** are traditional ways of describing the different arms of the immune system. Antibody, complement and other soluble molecules constitute the humoral effector systems, whereas T cells, NK cells, and phagocytes constitute the cellular effectors. It is now more useful to think of the systems that recognize free antigens and those that recognize cell-associated antigens. For example, cytotoxic T cells can recognize antigens presented on cell membranes, which have originated from within that cell, whereas antibody is important in the recognition of free, extracellular antigens.

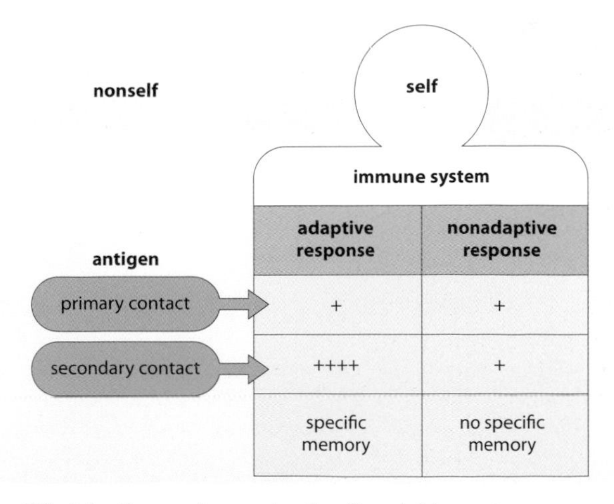

Fig. 3.2 Adaptive and nonadaptive (innate) immune responses.

ANTIBODY RESPONSE

After injection of an antigen, an antibody response develops, which may be divided into four phases: a lag phase in which no antibody is detected, followed by a phase in which the antibody titers rise logarithmically and then reach a plateau, and decline, as the antibodies are catabolized or cleared as immune complexes.

Primary and **Secondary antibody responses** The quality of the antibody response after the second (secondary) encounter with antigen differs from that after the first (primary) contact. The primary response has a longer lag phase, reaches a lower plateau, and declines more quickly than the secondary response. IgM is a major component of the primary response and is produced before IgG, whereas IgG is the main class represented in the secondary response. During their development, some B cells switch from IgM production to other classes, and this is the basis of the change in antibody isotype seen in the secondary response. Differences between the primary and secondary response are most noticeable when T-dependent antigens are used, but the route of entry and the way it is presented to T and B cells also affect the development of the response and the classes of antibody produced.

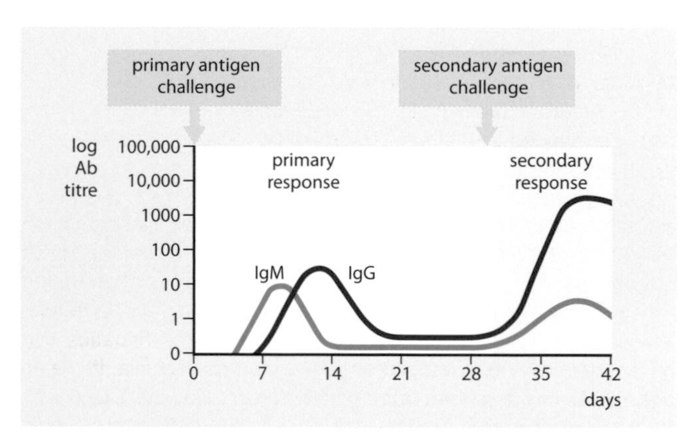

Fig. 3.3 Primary and secondary antibody responses.

Affinity maturation is the finding that the average affinity of the induced antibodies increases in the secondary response. The effect is largely confined to IgG and IgA, and is most marked when a low antigen dose is given in the secondary injection. Low levels of antigen bind preferentially to high-affinity B-cell clones and activate them—there is insufficient to activate low-affinity clones.

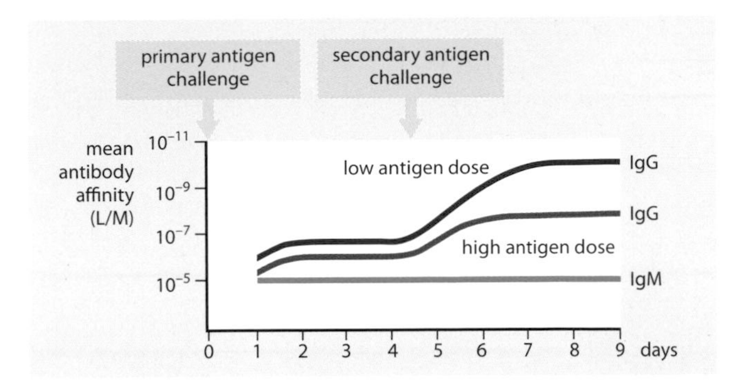

Fig. 3.4 Affinity maturation.

The underlying cellular basis of affinity maturation is the change in the affinity of B-cell clones, caused by somatic hypermutation of the antibody genes that occurs in germinal centers, where B cells compete for antigen on follicular dendritic cells. The process is accompanied by, but is not dependent on, class switching. It does not occur in the response to T-independent antigens, which is predominantly IgM antibody. Therefore the survival and development of high-affinity B cells depends on T cells.

Active immunization/Vaccination are the terms used for the active induction of protective immunity against a pathogen. This depends on the greater effectiveness of the secondary immune response. Vaccines may be live attenuated organisms, killed organisms, individual antigens of a pathogen, or modified antigens. In general, living organisms are more effective than killed ones or individual antigens, except in cases where the pathology is caused by a toxin (such as diphtheria). In this case a modified toxin or toxoid, which retains antigenicity but lacks pathogenicity, is preferred. Newer vaccines may be produced by genetic engineering. For example, genes for antigens of pathogenic viruses such as hepatitis can be inserted into nonpathogenic viruses such as vaccinia. It is also possible to insert antigen fragments that can stimulate T cells into such carrier viruses. For antigens that are only weakly immunogenic (such as some bacterial carbohydrates), coupling the antigen to an immunogenic carrier has often been successful. These preparations are called conjugate vaccines.

Passive immunization is the administration of antibodies preformed in another individual to contribute to protective immunity against a toxin. It is used when an individual's own active immune response would be too slow, for example in producing a response to a snake venom or tetanus toxin.

CELL COOPERATION

Cooperation between cells involved in immune responses occurs at many levels. Dendritic cells can take up antigen in the periphery and transport it to secondary lymphoid tissues (spleen, lymph nodes, etc.) for presentation to T cells. B cells and macrophages can also internalize antigen, process it and present it in association with MHC class II molecules to $CD4^+$ T_H cells. Cytokines produced by activated T_H2 cells stimulate B-cell growth and differentiation into plasma cells. Other cytokines can also activate T_C cells, APCs, and mononuclear phagocytes, thereby facilitating uptake of antigen. IgG antibodies can sensitize target cells for attack by NK cells. IgE antibodies can sensitize mast cells and basophils to release their inflammatory mediators when they bind specific antigen. Cytokines and antibodies are soluble mediators of cell cooperation, but leukocytes also signal directly via cell-surface receptors. The most important direct interaction involves MHC molecules/antigen peptides contacting the T-cell receptor, but other interactions are essential for cellular cooperation, including adhesion and co-stimulation.

Antigen presentation is the process by which antigen is presented to lymphocytes in a form they can recognize. Most $CD4^+$ T cells must be presented with antigen on MHC class II molecules, whereas $CD8^+$ T_C cells recognize antigen on class I MHC molecules. Antigen must be processed into peptide fragments before it can associate with MHC molecules. The way in which an antigen is processed and the type of MHC molecule it associates with determine which T cells will recognize it and whether the antigen is immunogenic or tolerogenic. It also affects the type of immune response generated.

Adhesion is an essential component of the interactions between leukocytes and other cells. It controls the position of the cell in lymphoid tissue, controls migration into tissues, and is a prerequisite for antigen presentation and many immune effector functions.

Co-stimulation Most immune responses are initiated by antigen triggering B or T cells. However, cellular activation also requires other signals. These may be delivered via co-stimulatory molecules (such as CD40 for B cells, CD28 for T cells) or by cytokines. This is sometimes called the two-signal hypothesis, in which antigen provides the first signal and the other co-stimulatory interactions provide the second signal. Cells that only receive a first signal may become anergic (tolerant) to their particular antigen.

Cytokines are signaling proteins, many of which are involved in signaling between cells of the immune system. The group includes the interleukins (IL-1 to IL-35), interferons (IFNs), tumor necrosis factors (TNFs), transforming growth factors (TGFs), and colony-stimulating factors (CSFs). The term lymphokines was originally used for those cytokines produced by lymphocytes.

T-cell help describes the cooperative interactions between TH2 cells and B cells in the production of the antibody response to T-dependent antigens, or between TH1 and TH17 cells and phagocytic cells in cell-mediated responses. In either case the antigen-presenting cell presents processed antigen to the T cell, receives co-stimulatory signals and is then triggered by specific cytokines. For example, a B cell internalizes its own specific antigen and presents it to the T cell. It transduces co-stimulatory signals via CD40 and is further activated by IL-4, IL-2, and IL-13.

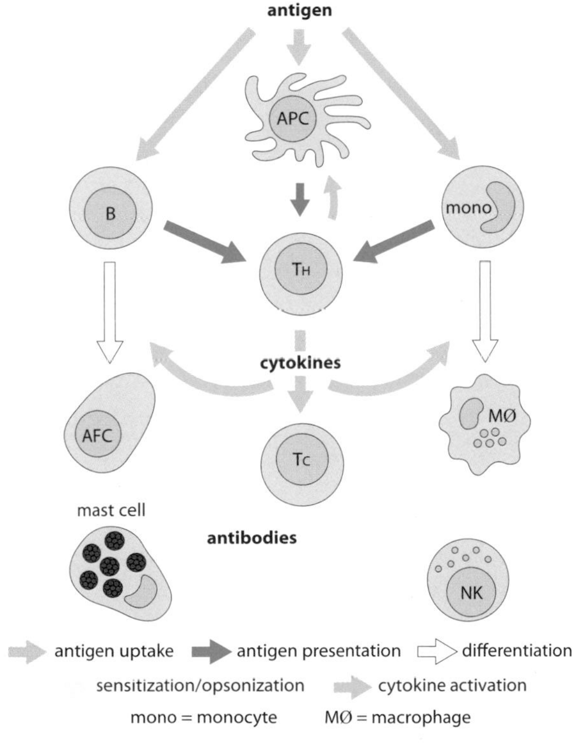

Fig. 3.5 Cooperation between cells in the immune response.

ANTIGEN PRESENTATION

Antigens are taken up by antigen-presenting cells in a variety of ways. B cells use surface antibody to bind and internalize their specific antigen. This is partly degraded (processed) and returned to the cell surface associated with MHC class II molecules, for recognition by TH2 cells. Theoretically, B cells can endocytose and present any antigen, but in practice they selectively concentrate only their own specific antigen in sufficient quantities. Mononuclear phagocytes phagocytose opsonized particles via their Fc and C3 receptors, which are then processed before presentation to TH1 cells. Immature dendritic cells take up antigen by phagocytosis using Fc, C3, scavenger, and lectin-family receptors. They lose these receptors and degrade antigen before migrating to lymph nodes, where they present it to T cells.

Antigen processing is the process of antigen breakdown and its association with MHC molecules. Blocking degradative pathways leaves cells unable to process and present most antigens. Different cell types have different capacities to degrade antigens and hence different abilities to stimulate T cells. There are two distinct pathways for antigen processing, used by MHC class I and II molecules. They are referred to as the internal and external pathways; MHC class I presents antigen from inside the cell, whereas MHC class II presents antigens that the cell has endocytosed.

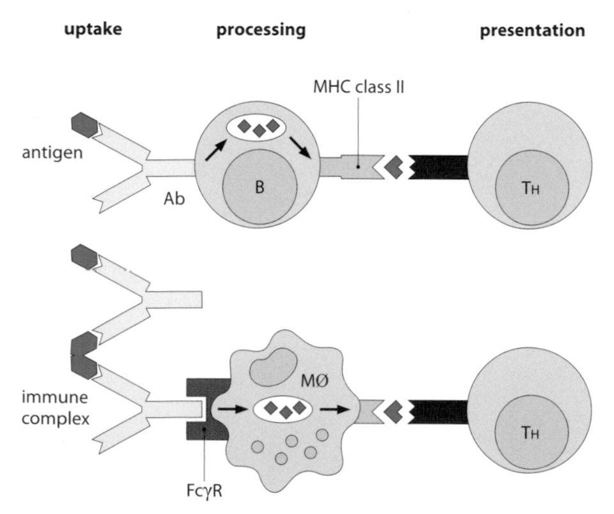

Fig. 3.6 Antigen processing and presentation by APCs.

Class II (external) pathway Antigens such as immune complexes that have been endocytosed by a cell associate preferentially with MHC class II molecules. They are partly degraded, and the endocytic vesicles containing peptide fragments then fuse with vesicles containing MHC class II molecules.

Invariant chain (Ii, CD74) MHC class II molecules are initially produced in association with an invariant chain, Ii, which is required for folding of the class II molecule and prevents peptides from binding to it in the endoplasmic reticulum. The invariant chain targets the class II molecules to the MIIC compartment.

MIIC compartment is an acidic endosomal compartment where antigenic peptide fragments and MHC class II molecules combine. The invariant chain is degraded, leaving a small peptide, CLIP, bound to the class II molecule. Once this has been replaced by an antigenic peptide, the class II:peptide complex can be finally processed (trimmed) before moving to the cell surface.

Antigenic peptides are the protein fragments that bind to MHC molecules. Class I molecules accommodate 8–9 amino acids in the peptide-binding groove, class II molecules 12–15 amino acids.

DM molecules are class II-like molecules that are required to facilitate loading of peptides onto the class II molecules.

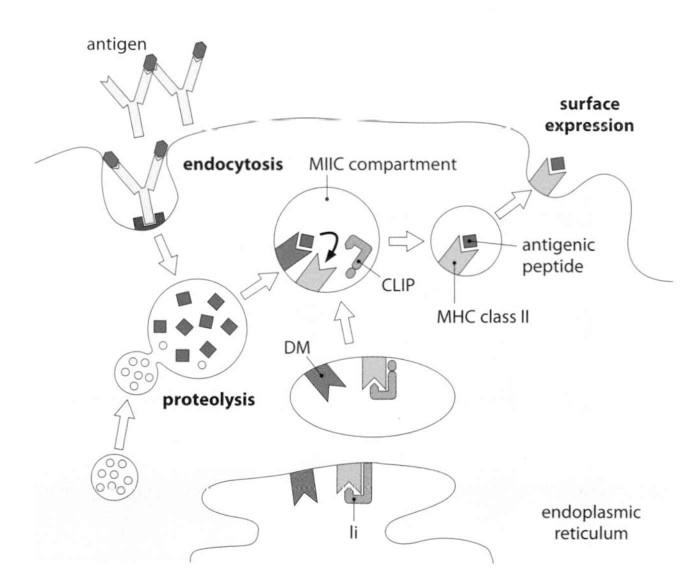

Fig. 3.7 Antigen presentation: MHC class II pathway.

Class I (Internal) pathway Antigens synthesized within a cell, such as viral polypeptides and the cell's own proteins, associate preferentially with MHC class I molecules. Peptide fragments from the cytosol are sampled and presented, for review by $CD8^+$ T cells.

Proteasomes are multicatalytic protease complexes that break down cytosolic proteins into fragments that may be loaded onto MHC class I molecules. Two components of the proteasome (LMP-2 and LMP-7) are encoded within the MHC.

TAP-1 and **TAP-2** are MHC-encoded members of the ABC family of transporters. They transfer peptides across the membrane of the endoplasmic reticulum to be loaded onto MHC class I molecules.

Calnexin is a molecular chaperone that stabilizes the class I α chain, until it associates with β_2-microglobulin and the peptide fragments. Once released from calnexin, assembly of the MHC:peptide complex takes place in a peptide loading complex and the peptide may then be trimmed by an ER-associated aminopeptidase. MHC:peptide complexes are transported to the cell surface, while incorrectly assembled complexes are degraded.

Anchor residues are critical amino acids that are required for an antigenic peptide to bind to an MHC molecule. The requirement for particular amino acids at each anchor position depends on the haplotype of the MHC molecule.

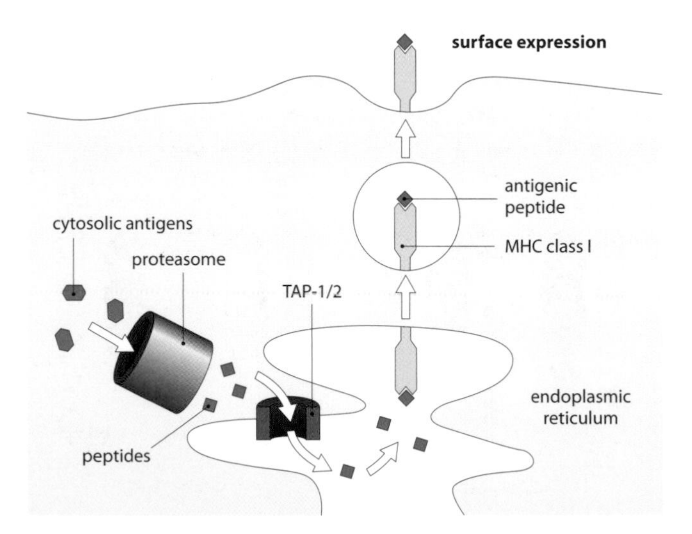

Fig. 3.8 Antigen presentation: MHC class I pathway.

MHC restriction describes the observation that T cells recognize antigen associated with particular MHC molecules, and usually do not recognize the same antigenic peptide if it associates with an MHC molecule of another haplotype. During development in the thymus, T cells that can interact with self MHC molecules are produced; these cells do not interact effectively with antigen-presenting cells of another MHC haplotype.

Cross-presentation may occur when an external antigen (normally presented by the class II pathway) is presented on MHC class I molecules. This mechanism can allow APCs to present viral antigens to $CD8^+$ cytotoxic T cells, even if they have not themselves become infected.

CD4 and **CD8** are functionally analogous molecules expressed on mature T cells; cells have either CD4 or CD8, but not both. CD8 consists of two disulfide-linked transmembrane polypeptides that can interact with the TCR on T cells and bind to a site in the α3 domain of class I molecules on the target cell (Figs 3.9 and 2.16). This interaction contributes to the stabilization of the immune recognition complex. CD4 has a single transmembrane polypeptide and binds MHC class II molecules on APCs.

lck is a kinase associated with CD4 and CD8. Binding of the T cell to MHC:antigen brings lck into proximity with the T cell receptor so that it phosphorylates CD3ζ to initiate T-cell activation.

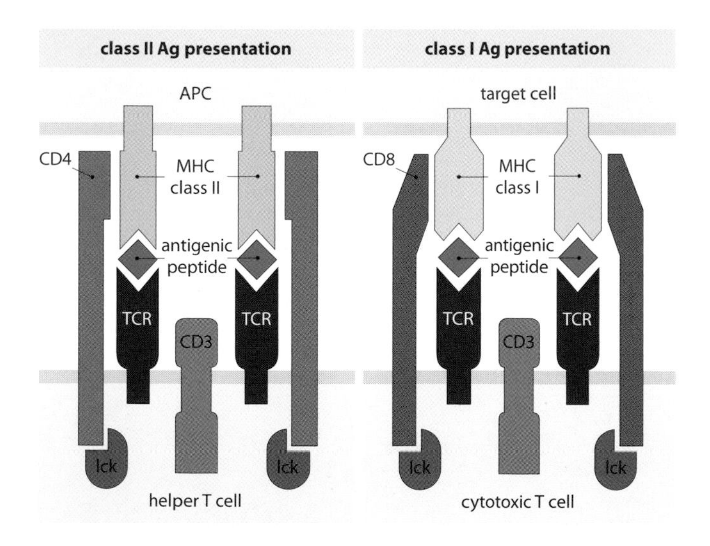

Fig. 3.9 Immune recognition by T cells.

T-CELL ACTIVATION

T cells require three types of signal for full activation:

- Antigenic peptide presented on an MHC molecule
- Co-stimulatory signals
- Signals from specific cytokines

If a cell does not receive a full set of signals, it will not divide and may even become anergic. Molecules such as CD2 and LFA-1 contribute to the adhesion between a T cell and an APC, and enhance activation signals, but co-stimulation transduced via CD28 is essential for activation.

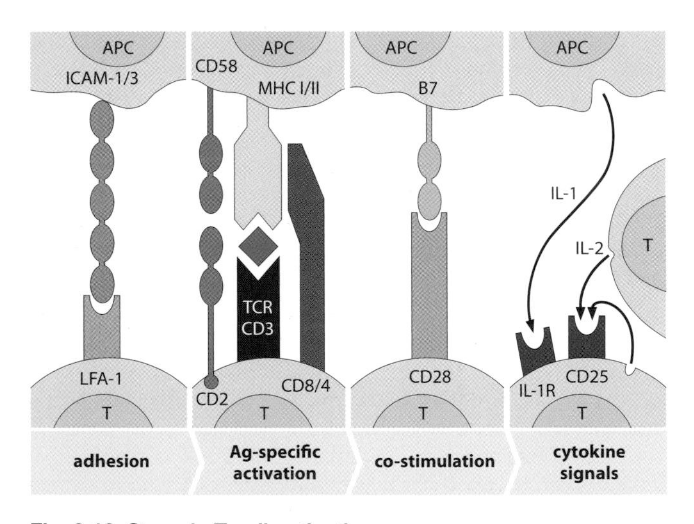

Fig. 3.10 Steps in T-cell activation.

Lymphocyte functional antigen-1 (LFA-1, CD11a/CD18) is a member of the β2-integrin family present on most leukocytes. It consists of two polypeptides (CD11a and CD18) that interact with intercellular adhesion molecules, ICAM-1, ICAM-2, and ICAM-3. Transient adhesion between lymphocytes and APCs is mediated by LFA-1 binding to ICAM-1 and ICAM-3. Lymphocyte activation enhances the affinity of LFA-1, thereby extending the interaction time between the T cell and APC. Binding of LFA-1 with ICAM-1 and ICAM-2 is also important in the attachment of leukocytes to endothelium and in the migration of cells across endothelium in normal tissues and at sites of inflammation.

ICAM-3 (CD50) is an adhesion molecule present on many leukocytes, which increases after lymphocyte activation and contributes to T-cell interactions with APCs.

CD2 (LFA-2) and **CD58 (LFA-3)** are a pair of molecules involved in T-cell activation. CD2 is expressed on all T cells. It has a single transmembrane polypeptide that acts as a receptor for CD58, a molecule that is widely distributed on many cell types. Interaction of CD2 with CD58 enhances the binding of the T cell to its target, amplifying the activation signal initiated by the TCR:CD3 complex.

CD28 and **B7 (CD80, CD86)** are molecules that critically regulate T-cell activation. CD28 is a co-stimulatory receptor present on 80% of $CD4^+$ T cells and ~50% of $CD8^+$ cells. The molecules B7-1 (CD80) and B7-2 (CD86), expressed on many APCs, are the principal ligands for CD28. As an immunological synapse forms, CD28 is released from intracellular stores, where it enhances the initial weak signal from the TCR. The cytoplasmic portion of CD28 associates with phosphatidylinositol 3-kinase which, in association with signals from the TCR, activates the MAP-kinase signaling pathway.

B7-1 (CD80) and **B7-2 (CD86)** are constitutively expressed on dendritic cells and most mononuclear phagocytes; expression is enhanced by GM-CSF, IFN-γ, and ligation of TLRs (for example, by LPS). B7 is induced on B cells by antigen binding, LPS stimulation, and ligation of CD40.

Immunological synapse is the complex of interacting molecules that link the APC and the T cell. Initially, adhesion molecules (LFA-1/ICAM-1, etc.) allow the cells to adhere to each other. As MHC molecules on the APC start to interact with the TCR complex, the adhesion molecules are relegated to the outside of the synapse, the pSMAC (peripheral supramolecular activation complex), while the TCR, CD2/CD58, CD28/B7, and MHC molecules localize at the center of the synapse—the cSMAC (Fig. 3.11).

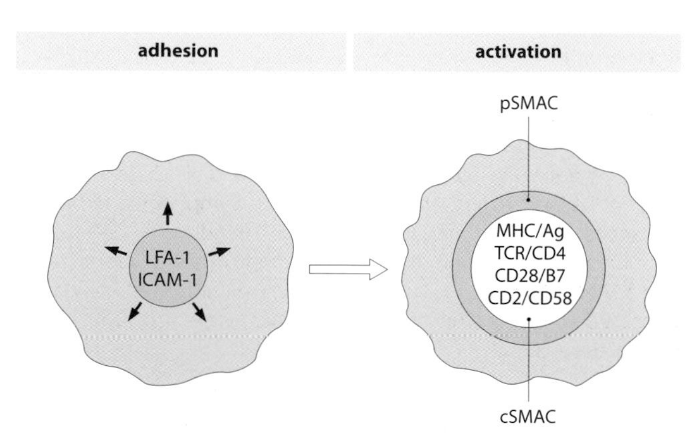

Fig. 3.11 Formation of an immunological synapse.

IL-2 receptor (IL-2R, CD25) is induced on activated T cells. The high-affinity receptor is formed when the induced α chain (CD25) associates with β and γ chains (CD122, CD132), which together form the low-affinity receptor. IL-2 is essential for T-cell division, and the high-affinity receptor persists for several days after T-cell activation. CD25 is also a characteristic marker of naturally occurring regulatory T cells (TREG cells), which may act by mopping up excessive IL-2, limiting activation of antigen-stimulated T cells.

CTLA-4 (CD152) is an alternative ligand for B7, which is not expressed on resting T cells but is induced after T-cell activation as CD28 declines. CTLA-4 has a higher affinity for B7 than does CD28, and by competing with CD28 for B7, CTLA-4 counters the co-stimulatory action of CD28. CTLA-4 is also constitutionally expressed in TREG cells. Mice deficient in CTLA-4 are more susceptible to autoimmune diseases; this is thought to be due both to excessive T-cell activation and reduced control by TREG cells.

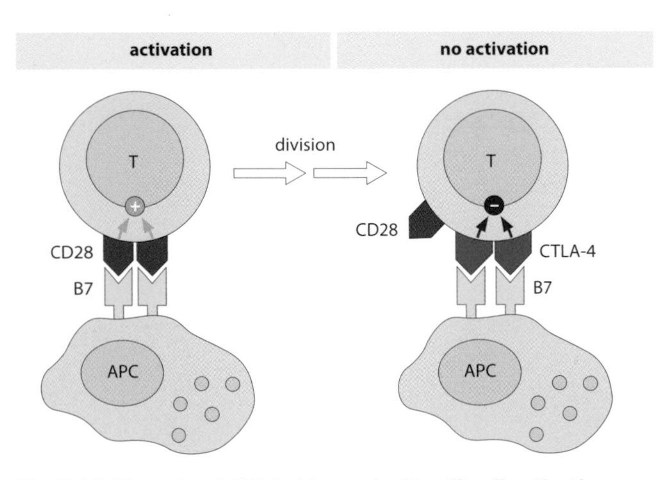

Fig. 3.12 The role of CTLA-4 in controlling T-cell activation.

PD1 (CD279) (Programmed Death-1) is another inhibitory receptor, belonging to the same family as CD28 and CTLA-4. It is expressed late after T-cell activation, and can bind to its ligands PD-L1 (CD273) or PD-L2 (CD274), which belong to the B7 family; the ligands are expressed on antigen-presenting cells. PD1 is also present on B cells, dendritic cells, and monocytes. It is thought to limit T-cell activation and to help prevent autoimmunity. In humans, polymorphisms in PD1 are linked to rheumatoid arthritis, Graves' disease, type I diabetes, and multiple sclerosis.

CYTOKINE RECEPTORS

Cytokine receptors determine the responsiveness of a cell to particular cytokines. Receptors for IL-1, TNF, and the interferons are widely distributed. Others are induced on particular lineages for limited periods. For example, the high-affinity IL-2 receptor is present on antigen-activated cells for a limited period, but expression wanes if the T cell is not re-stimulated with antigen. Expression of IL-4 receptors occurs on activated B cells in an analogous fashion. Receptors for colony-stimulating factors appear during hemopoietic cell differentiation on the appropriate developing cells (see Fig. 1.11). The cytokine receptors fall into families on the basis of structural motifs and shared chains. For example, the receptors for IL-2, IL-4, IL-7, IL-9, and IL-15 have a common signaling polypeptide (CD122), but individual cytokine-binding chains. IL-3 and IL-5 share a different chain.

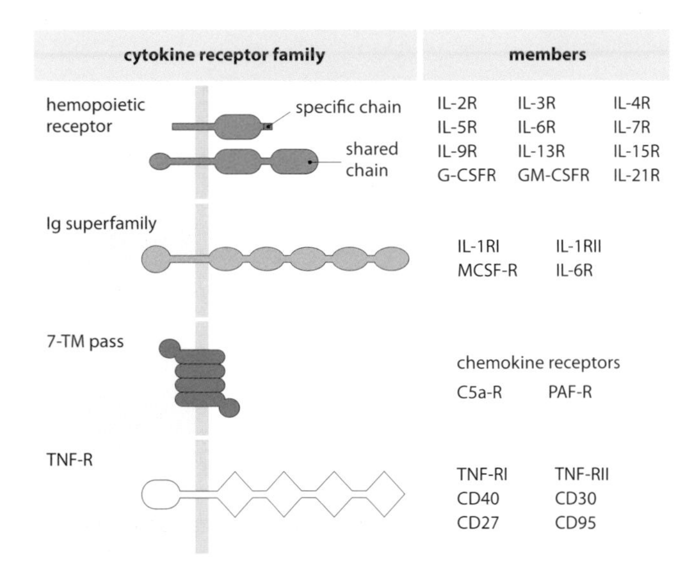

Fig. 3.13 Families of cytokine receptors.

Soluble cytokine receptors and cytokine inhibitors Several cytokine receptors are produced in a soluble, truncated form, lacking the membrane-spanning domains. Examples are the soluble TNF-R, IFN-γR, and IL-1R. They are thought to limit the effects and zone of action of cytokines *in vivo*. Cytokine inhibitors have also been identified. For example, IL-1RA (IL-1 receptor antagonist) binds to the IL-1 receptor but does not activate the cell.

B-CELL ACTIVATION

B cells responding to T-dependent antigens require three types of signal for their activation. The first signal is mediated by the binding of antigen, which is internalized, processed, and presented to T cells. Then a co-stimulatory signal is transduced via CD40, which is bound by CD40L on the T cells. Thereafter B cell division, differentiation, and Ig class switching are driven by a large number of different cytokines. Type 2 T-independent antigens, such as polysaccharides that cross-link B-cell-surface antibody, can activate B cells directly, although such cells still need cytokine signals.

Intermolecular help refers to the way in which B cells taking up particles with several different antigens (such as a virus) can then present all of those antigens to T cells. They thus get help from T cells recognizing antigens that they themselves do not recognize.

CD40 is a surface receptor on B cells, follicular dendritic cells, dendritic cells, macrophages, endothelium, and hemopoietic progenitors. It belongs to the TNF receptor family. It provides a critical co-stimulatory signal to B cells that is also needed for the development of germinal centers and B-cell memory.

CD40L (CD154) is the ligand for CD40, induced transiently on $CD4^+$ T cells and some $CD8^+$ cells, after activation. It is also present on eosinophils and basophils. CD40L is essential for the delivery of T-cell help to B cells. A defect in CD40L causes impaired class switching and results in hyper-IgM syndrome.

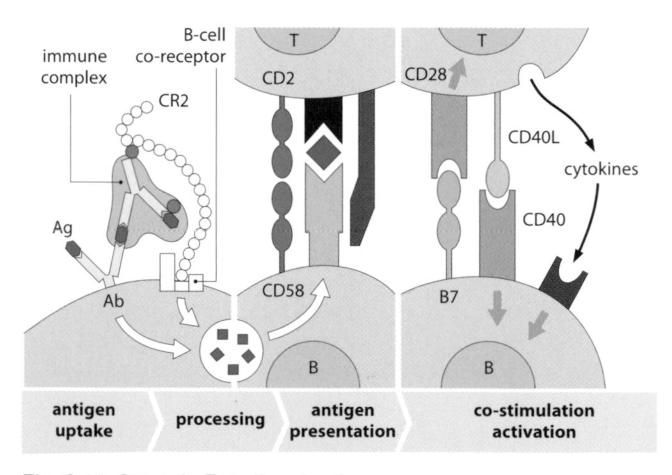

Fig. 3.14 Steps in B-cell activation.

CD72 and **CD100** are co-stimulatory molecules for B-cell activation. CD72 on B cells binds to CD100, a widely distributed member of the semaphorin family, enhancing activation mediated by CD40.

CD45 (Leukocyte common antigen) is a phosphatase present on all leukocytes, and is produced in six different forms, using combinations of exons. B cells express the largest variant of CD45. It controls lymphocyte activation by acting on lck, which can phosphorylate the signaling portion of the TCR (CD3) and BCR (CD79).

B-cell co-receptor complex (CD19, CD21/CR2, CD81/TAPA-1) amplifies signaling from the B cell antigen receptor. Cross-linking of CD19 to surface Ig makes a B cell 100 times more sensitive to antigen. This is important in the initial development of an antibody response when B-cell antibody affinity is low. Immune complexes formed in the primary immune response may fix complement C3 and then bind to CD21 on the B cell, which is complement receptor type 2 (CR2). If the complexed antigen is recognized by the B-cell receptor, the complex cross-links the co-receptor complex and surface Ig, thereby activating the B cell very efficiently. This may explain the observation that complement is required for the development of secondary antibody responses and B-cell memory.

CD23 (FcεRII) is a low-affinity IgE receptor with a lectin domain that also binds CR2. It is expressed on B cells, activated macrophages, and follicular dendritic cells, but may also be released in a soluble form to act as a B-cell co-stimulatory factor.

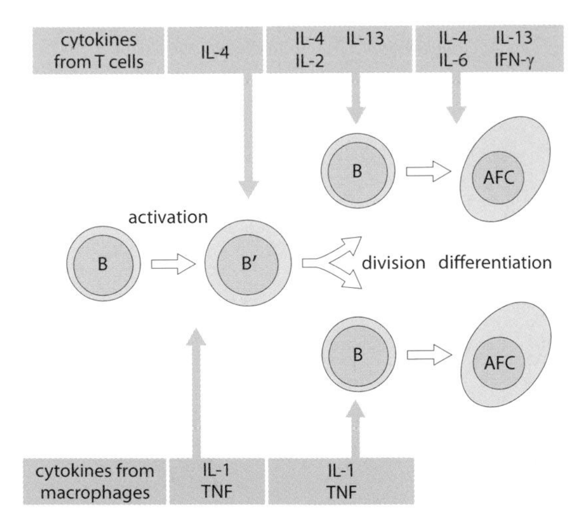

Fig. 3.15 Role of cytokines in B-cell development.

CYTOKINES

Cytokines, released by leukocytes and sometimes other cells, are very important in controlling the development of immune responses. They modulate the differentiation and division of hemopoietic stem cells and the activation of lymphocytes and phagocytes. They control the balance between cell-mediated responses and antibody production. Others can act as mediators of inflammation or as cytotoxins. Many cytokines have more than one action (pleiotropy), and different cells produce distinct blends of cytokines. The ability to respond to a cytokine depends on the expression of a specific receptor. Often more than one cytokine signal is required for a response, and in this case the different cytokines act synergistically. Helper T cells are particularly important sources of cytokines. Most cytokines act on cells other than those that produced them (paracrine action), but some can also stimulate the cell that produced them (autocrine action).

JAKS and **STATs** Cytokines signal cell activation by binding to specific receptors (see p. 65) that activate intracellular signaling pathways. Receptors that belong to the hemopoietic cytokine receptor family are associated with Janus kinases (JAKs). When the receptors become clustered after cytokine binding, the JAKs phosphorylate STATs (signal transducers and activators of transcription). The activated STATs, in association with other proteins, form transcription factors that migrate to the nucleus, bind to gene promoters, and induce the sets of genes that are associated with the response to each of the cytokines. Different JAKs and STATs are used by different cytokines and their receptors. In the example below, the interferon-α receptor is associated with JAKs Tyk2 and Jak1. These phosphorylate STAT1 and STAT2, which associate with p48 to form a transcription factor.

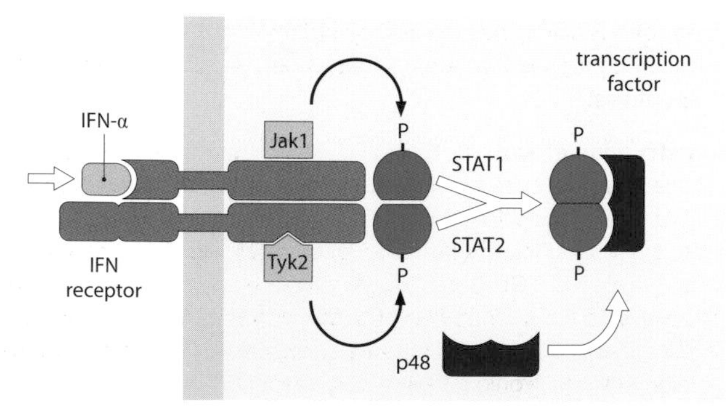

Fig. 3.16 JAKs and STATs in cytokine receptor signaling.

Interferon-γ (IFN-γ) is released by antigen-activated TH1 cells. In addition to its antiviral effects, IFN-γ enhances MHC class I on many cells and increases MHC class II and B7 expression on B cells and macrophages, thereby enhancing antigen presentation. At high levels it can induce class II on some tissue cells. It increases IL-2 receptors on Tc cells, enhances the cytotoxic activity of NK cells and promotes B-cell differentiation. IFN-γ also acts as a macrophage arming factor, increasing Fc receptor expression, the respiratory burst, and nitric oxide production, so enhancing the ability of macrophages to destroy pathogens. It also inhibits TH2 and TH17 cells, and so reinforces TH1-type immune responses.

Migration inhibition factor (MIF) is released by activated T cells and inhibits macrophage migration. It causes the accumulation and activation of macrophages at the site of inflammation and is elevated in many chronic inflammatory conditions. MIF binds to CD74, acting on various transcription factors and modulating the cell damage sensor/tumor suppressor p53.

Tumor necrosis factor (TNF) and **Lymphotoxin (LT)** are structurally related cytokines encoded within the MHC. Lymphotoxin, released by Tc cells, is also called TNF-β, whereas the original TNF, released by macrophages and several other cells, is TNF-α. A transmembrane form of lymphotoxin (LT-β), also produced by T cells, trimerizes with LT-α. TNF enhances the adhesiveness of vascular endothelium for leukocytes by inducing E-selectin, VCAM-1 and ICAM-1, thus promoting transendothelial migration. TNF also causes the mobilization of fat, which is partly responsible for the wasting (cachexia) seen in some chronic diseases. It also synergizes with IFN-γ in many of its actions such as MHC induction and macrophage activation. TNF and lymphotoxin can also induce cell death by apoptosis. Of the three receptors for this group of cytokines, one (TNFR1) has an intracytoplasmic death domain, which can recruit proteins that activate caspases, the principal mediators of apoptosis.

Transforming growth factor-β (TGF-β) is a group of five cytokines released by many cell types, including macrophages and platelets. They are mitogenic for fibroblasts and some other mesenchymal cells, and they enhance the production of extracellular matrix proteins. In general, TGF-β is strongly inhibitory of immune responses as it prevents the proliferation of both T and B cells, and it seems to be essential in controlling immune reactivity—TGF knockout mice develop severe chronic inflammatory reactions.

Interleukins (IL-1 to IL-35) are a diverse group of cytokines; most newly discovered cytokines are placed into this series. The functions are outlined in Fig. 3.17. Many of the interleukins fall into structurally related families:

IL-1 family—IL-1, IL-18, IL-33
IL-2 family—IL-2, IL-12, IL-15
IL-10 family—IL-10,IL-19, IL-20, IL-22, IL-24, IL-26
IL-12 family—IL-12, IL-23, IL-27, IL-35

cytokine	source	target	principal effects
IL-1α	macrophage fibroblast lymphocytes	lymphocytes macrophages endothelium	lymphocyte co-stimulation phagocyte activation ↑ endothelial adhesion molecules
IL-1α	epithelial cells astrocytes	other	induced fever and sleep ↑ prostaglandin synthesis
IL-2	T cells	T cells NK cells B cells	T-cell growth and activation NK-cell activation and division
IL-3	T cells thymic epithelium	stem cells	multilineage hemopoietic factor
IL-4	TH2 cells bone marrow stroma	B cells	activation and division promotes class switch → IgG1 and IgE
IL-5	TH2 cells	eosinophils B cells	development and differentiation
IL-6	macrophages endothelium TH2 cells	T cells B cells hepatocytes	lymphocyte growth B-cell differentiation acute-phase protein synthesis
IL-7	bone marrow stroma	pre-B cells pre-T cells	division
IL-8 (CXCL8)	endothelium monocytes fibroblasts	neutrophils monocytes T cells	activation/chemotaxis
IL-9	$CD4^+$ T cells	T cells mast cells	division promotes development
IL-10	TH2 cells	TH1 cells	inhibits cytokine synthesis
IL-11	bone marrow stroma	stem cells plasma cells	division proliferation
IL-12	B cells macrophages	TH0 cells NK cells	TH1-cell development activation
IL-13	TH2 cells	B cells macrophages	division and differentiation ↓cytokine production

cytokine	source	target	principal effects
IL-14	T cells	B cells	proliferation ↓ Ig synthesis
IL-15	monocytes	T cells B cells	division
IL-16	$CD8^+$ T cells	$CD4^+$ T cells	chemotactic
IL-17	TH17 cells	many cells	proinflammatory
IL-18	macrophages keratinocytes	blood mononuclear cells	induces IFN-γ and NK cell activity
IL-19	B cells monocytes	mononuclear phagocytes	induces IL-6 and TNF-α
IL-20	skin	keratinocytes	keratin synthesis
IL-21	T cells mast cells	B cells T cells NK cells	co-stimulates B and T cells NK proliferation and maturation
IL-22	T cells	liver	acute-phase protein synthesis
IL-23	dendritic cells macrophages	memory T cells dendritic cells	TH17 differentiation antigen presentation
IL-24	blood mononuclear cells	tumor cells	apoptosis inhibition
IL-25	TH2 cells	mucosal epithelia	eosinophilia
IL-26	TH17 cells	epithelial cells	induces ICAM-1
IL-27 (IL-30)	dendritic cells APCs	B cells T cells hemopoietic stem cell	regulates inflammation TH1 differentiation
IL-28/ IL-29	TREG cells immature DCs	keratinocytes melanocytes	induce antiviral state
IL-31	TH2 cells	epithelial cells keratinocytes	proinflammatory
IL-32	monocytes macrophages	mononuclear phagocytes	induces TNF, CXCL8, CXCL2 promotes differentiation
IL-33	endothelium epithelium	T cells mast cells basophils	induces TH2 cytokines
IL-34	tissue cells	monocytes	differentiation
IL-35	TREG cells	T cells	suppresses TH17 cells proliferation of TREG cells

Fig. 3.17 The interleukins.

PHAGOCYTOSIS

Phagocytosis/Endocytosis is the process by which cells engulf particles and microorganisms. The particles first attach to the cell membrane of the phagocytic cell, either by general receptors, such as the mannose receptor, which binds bacterial carbohydrates, or by receptors for opsonins, such as IgG or C3b. Then the cell extends pseudopodia around the particle and internalizes it. Antibacterial, oxygen-dependent killing mechanisms are activated and lysosomes fuse with the phagosome. The lysosomal enzymes damage and digest the phagocytosed material, and digestion products are finally released. Endocytosis is a term that includes phagocytosis and pinocytosis (internalization of fluid).

Opsonization occurs when particles, microorganisms, or immune complexes become coated with molecules that allow them to bind to receptors on phagocytes, thereby enhancing their uptake.

Opsonins are molecules that bind to particles to be phagocytosed and to receptors on phagocytes, so acting as an adaptor between the two, such as IgG, C3b, C-reactive protein.

Immune adherence, effected by IgG and C3 products, refers to the attachment of opsonized particles to phagocytes, by binding to Fc and complement receptors (see pp. 74 and 75).

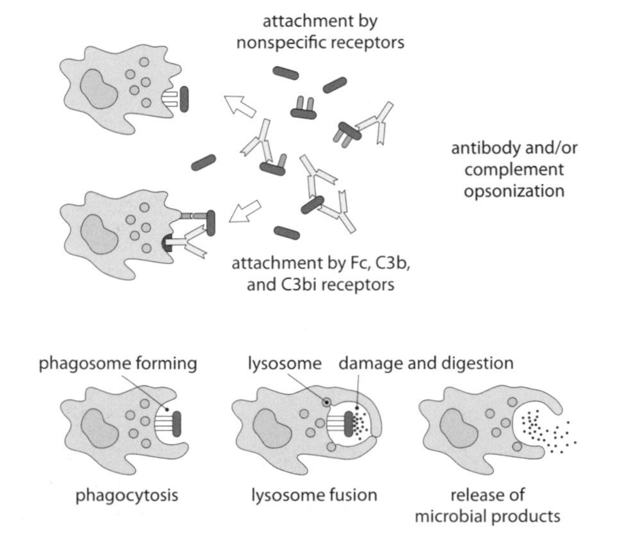

Fig. 3.18 Stages of phagocytosis.

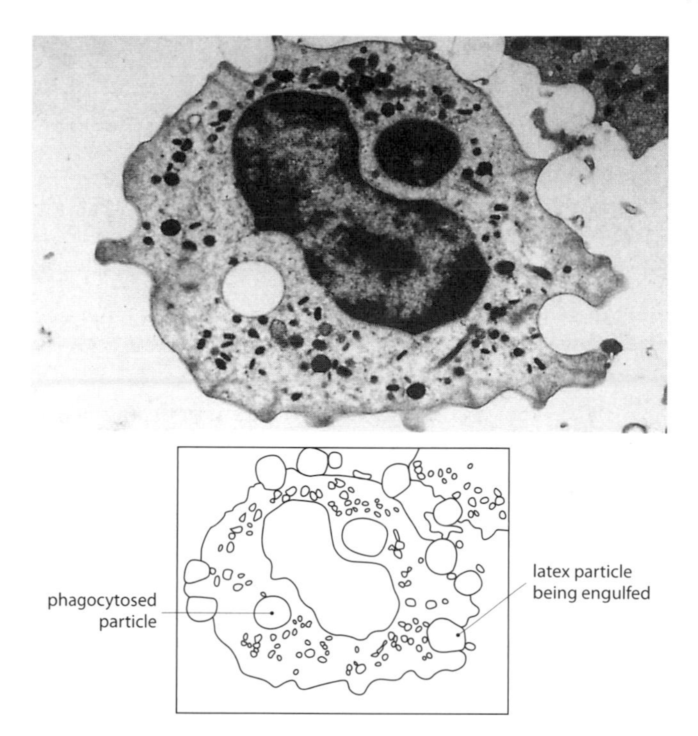

Fig. 3.19 Phagocytosis of latex by macrophages.

Phagosomes are membrane-enclosed intracellular vesicles that contain phagocytosed materials.

Lysosomes are organelles present in all cells. They contain enzymes that, in macrophages, damage and digest the phagocytosed material. Newly formed lysosomes are called 'primary', and mature lysosomes are 'secondary.'

Phagolysosomes are formed by the fusion of phagosomes and lysosomes. Immediately after fusion, there is a brief rise in the pH of the phagolysosome, when neutral proteases (such as collagenase, elastase) and cationic proteins are active. Subsequently the pH falls and acid proteases (such as glycosidase, lipase) become active.

Frustrated phagocytosis occurs when phagocytes attach to material that cannot be phagocytosed (such as a basement membrane). The cells may release their lysosomal enzymes to the exterior (exocytosis). This process is thought to cause some of the damage in immune complex disease.

COMPLEMENT RECEPTORS

There are four different kinds of receptor for C3b or iC3b (CR1 to CR4), and three of them act as opsonic receptors for immune complexes on cells of the mononuclear phagocyte lineage.

CR1 (CD35) is a transmembrane protein consisting of a single polypeptide that is expressed on phagocytic cells, where it acts as a receptor for immune complexes. On human erythrocytes it facilitates transport of complexes to phagocytic cells in the spleen and liver. On other cells its principal function is to act as a cofactor for factor I.

CR2 (CD21) is structurally similar to CR1. It forms part of the B-cell co-receptor complex (Fig. 3.14) and is also present on follicular dendritic cells. It is involved in the uptake of immune complexes to germinal centers and in the development of B-cell memory.

CR3 (CD11b/CD18) is an integrin expressed on mononuclear phagocytes, neutrophils, and NK cells, where it facilitates the uptake of immune complexes with bound C3d. It is also involved in monocyte migration into tissues, by binding to ICAM-1.

CR4 (CD11c/CD18, p150/95) is an integrin that shares a β chain with CR3 and LFA-1. It has similar functions to CR3 and is highly expressed on tissue macrophages and dendritic cells.

CD93, present on monocytes, neutrophils, endothelium, and activated macrophages, was originally identified as a C1q receptor (C1qRp), but is now thought to be an adhesion molecule involved in clearance of apoptotic cells and in antimicrobial defense.

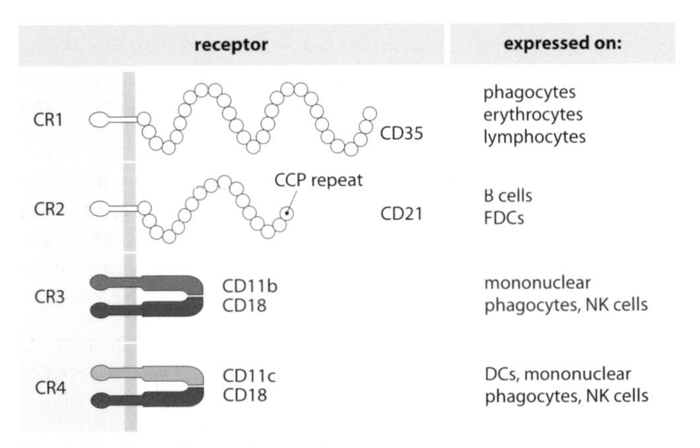

Fig. 3.20 Complement receptors.

Fc RECEPTORS

There are three well-defined receptors for IgG on phagocytes, which facilitate the uptake of immune complexes and allow cytotoxic cells to interact with targets. Two receptors for IgE have been described, FcεR1 and FcεRII; the first has a role in the control of inflammation, and the second has a role in immunoregulation and defense against parasitic worms.

FcγRI (CD64) is a high-affinity IgG receptor, capable of binding monomeric antibody. It is a characteristic marker of mononuclear phagocytes, but may also be expressed on neutrophils. It is involved in the uptake of immune complexes.

FcγRII (CD32) is a low-affinity receptor present on mononuclear phagocytes, neutrophils, eosinophils, platelets, and B cells. On phagocytes it facilitates uptake of large immune complexes, but on B cells it is thought to be involved in the control of antibody production. Cross-linking of the surface antibody (BCR) and FcγRII receptors on B cells leads to suppression of the B cell. Activation of platelets by immune complexes bound to their Fc receptors can cause degranulation with release of inflammatory mediators.

FcγRIII (CD16) is a low-affinity IgG receptor that occurs in two forms. On NK cells it is a transmembrane glycoprotein (FcγRIIIa) that can cross-link them to target cells sensitized with antibody. Engagement of this receptor on NK cells leads to cell activation. On macrophages and neutrophils, FcγRIII is a GPI-linked receptor (FcγRIIIb) attached to the outer leaflet of the plasma membrane, where it can bind immune complexes but cannot signal.

FcεRI is a high-affinity IgE receptor found on mast cells and basophils. These cells are sensitized by monomeric IgE bound to the receptor. When the specific antigen cross-links IgE bound to these receptors, it causes degranulation with release of histamine and other inflammatory mediators.

FcεRII (CD23) is a low-affinity IgE receptor with an immunoregulatory function present on some B cells. A soluble form of the receptor acts as a signaling molecule between lymphocytes (see p. 67). It is also present on eosinophils, where it may allow them to engage parasites (such as schistosomes) coated with IgE.

FcαR (CD89) is expressed on phagocytic cells, and on some B and T cells, particularly in Peyer's patches and the lamina propria. Hence it seems to be involved in the regulation of IgA synthesis.

PHAGOCYTE MICROBICIDAL SYSTEMS

Respiratory burst Shortly after phagocytosing material, neutrophils and macrophages undergo a burst of activity, during which they increase their oxygen consumption. This is associated with increased activity of the hexose monophosphate shunt and the production of H_2O_2 and $O_2^{\bullet}$.

Oxygen-dependent killing occurs within phagosomes and is activated via cross-linking of the phagocytes' C3 and Fc receptors. Initially an enzyme, NADPH oxidase, is assembled in the phagosome membrane; it reduces oxygen to superoxide ($^{\bullet}O_2^-$), which can then give rise to hydroxyl radicals ($^{\bullet}HO$), singlet oxygen ($\Delta g'O_2$) and hydrogen peroxide (H_2O_2).

Reactive oxygen intermediates (ROIs) are the labile products of the oxygen-dependent killing pathway (Fig. 3.21) and can damage endocytosed bacteria. Cells prevent damage to themselves, by redox pathways involving the tripeptide glutathione, but some bacteria deploy similar defenses against ROIs.

Myeloperoxidase present in lysosomes can enter the phagosome where, in the presence of H_2O_2, it converts halide ions into toxic halogen compounds (such as hypohalite). Endocytosed peroxidase or catalase can also perform this reaction.

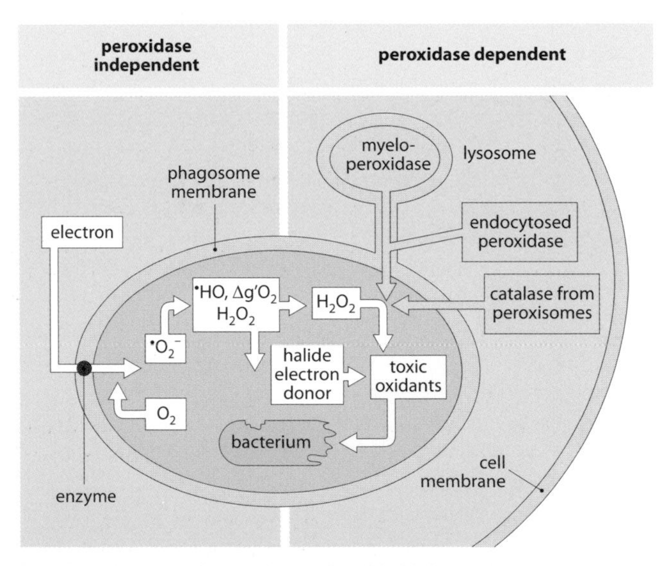

Fig. 3.21 Oxygen-dependent microbicidal activity.
$\Delta g'O_2$ = singlet oxygen $^{\bullet}HO$ = hydroxyl radical

Reactive nitrogen intermediates (RNIs) Murine macrophages that have been activated by IFN-γ and triggered with TNF express inducible nitric oxide synthetase (iNOS), which catalyses the production of nitric oxide, NO, which is toxic for some bacterial and fungal pathogens. Although human macrophages do not produce much NO, other cells, such as neutrophils, can do so. NO combines with ROIs to produce cytotoxic peroxynitrites.

Granules are specialized lysosomes of granulocytes that contain various bactericidal proteins. For example, neutrophil myeloperoxidase is in the primary (azurophilic) granules, whereas lactoferrin is in the secondary (neutrophil-specific) granules. Granule and lysosome contents are listed below.

Cationic proteins, found in neutrophil granules and in some macrophages, damage the outer phospholipid bilayer of some Gram-negative bacteria under alkaline conditions. This activity is produced by a number of molecules, including defensins and cathelicidins; some (such as cathepsin G) are enzymatically active.

Defensins are a group of small antimicrobial cytotoxic peptides that can be subdivided into three families, α, β, and θ. α-Defensins are found in the granules of neutrophils and macrophages in several species including humans. β-Defensins are present in the neutrophil granules of all mammals and in some epithelial cells. θ–Defensins are confined to primate granulocytes. Defensins are cationic proteins with a wide spectrum of antibacterial and antifungal actions that act by selectively damaging membranes with low levels of cholesterol and a high proportion of negatively charged phospholipids. They have some structural similarities to chemokines and have additional roles in opsonization and chemotaxis, by acting on chemokine receptors. For example, the defensin HBD-2 resembles CCL20, and both bind to the chemokine receptor CCR6.

Cathelicidins are a diverse family of small polypeptides with a common cathelin domain that are stored within granules of myeloid cells. On activation the cathelin domain is enzymatically removed and the peptides are released. In addition to antimicrobial properties, some cathelicidins have chemotactic and angiogenic activity and promote wound healing.

Lactoferrin is found in neutrophil granules. It binds tightly to iron, and deprives bacteria of this essential nutrient. Neutrophils loaded with iron are inefficient at destroying bacteria.

Lysozyme (muramidase) is an enzyme that digests a bond in the cell-wall proteoglycan of some Gram-positive bacteria. It is secreted constitutively by neutrophils and some macrophages and is present in many of the body's secretions.

Macrophage activation refers to the enhanced microbicidal (or anti-tumor) activity seen in response to stimulation by inflammatory cytokines (TNF-α, IL-1, IFN-γ), complement fragments, and bacterial products that activate the Toll-like receptors. Activated cells secrete more enzymes and produce more superoxide and RNIs due to inducible nitric oxide synthetase. Figure 3.22 shows that macrophages treated with IFN-γ (left) have a greater capacity to destroy the parasite *Leishmania donovani* than do untreated macrophages (right).

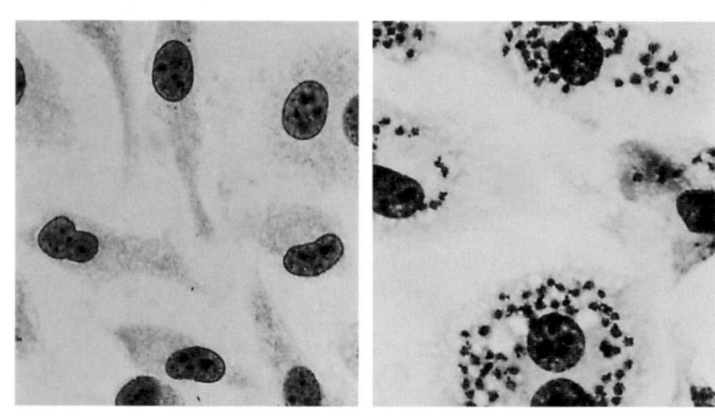

Fig. 3.22 Microbicidal activity of IFN-γ-activated macrophages.

Macrophage activation also induces expression of MHC class II and B7 to enhance antigen presentation. Enhanced phagocytosis of activated macrophages is related to increased expression of Fc and C3 receptors. Some receptors for chemotactic molecules (such as C5aR) are reduced; others are increased (such as CXCR3).

Macrophages can also be activated by cytokines released by TH2 cells, including IL-4 and IL-13. Such 'alternatively activated' macrophages increase expression of the mannose receptor and MHC class II, but do not show increased microbicidal activity.

nRAMP (natural resistance associated macrophage protein) is an ion pump that removes divalent cations from the phagosome, increasing macrophage resistance to mycobacterial infection.

Metalloproteases (MMP and **ADAM)** are zinc-containing enzymes involved in the degradation of extracellular matrix (matrix metalloproteases, MMP). ADAMs are transmembrane proteins containing a disintegrin and metalloprotease domain that modulate cell adhesion. Macrophage activation causes the synthesis of a number of new MMPs that are involved in tissue remodeling.

INTRACELLULAR RECEPTORS FOR PATHOGENS

Macrophages have a number of cytosolic molecules that can recognize intracellular bacterial and viral infections:

NOD-like receptors (NLRs) including NOD1 (nucleotide-binding oligomerization domain-containing protein-1) and NOD2 recognize bacterial peptidoglycans, for example from *Salmonella* and *Shigella*. Binding of peptidoglycans causes activation of NFκB and MAP-kinase pathways to induce transcription of cytokines that control inflammation.

RIG-like receptors (RLRS) include RIG-1 (retinoic acid inducible gene-1), which recognizes short dsRNA, and MDA5, which recognizes long dsRNA; dsRNA may be produced during viral replication. Binding of these receptors induces activation of NFκB. Both NLRs and RLRs are components of inflammasomes.

Inflammasomes are multicomponent complexes produced in myeloid cells, that include caspase-1 (= interleukin-1 converting enzyme, ICE). ICE cleaves pro-IL-1β and pro-IL-18 into their active forms, which promotes inflammation. The precise composition of the inflammasome depends on the inducing agent (NLR, RLR, etc.) Assembly of the inflammasome also activates caspases to cause cell death by pyropoptosis.

Pyropoptosis describes programmed cell death after activation of inflammosomes, with the release of the pro-inflammatory cytokines IL-1 and IL-18.

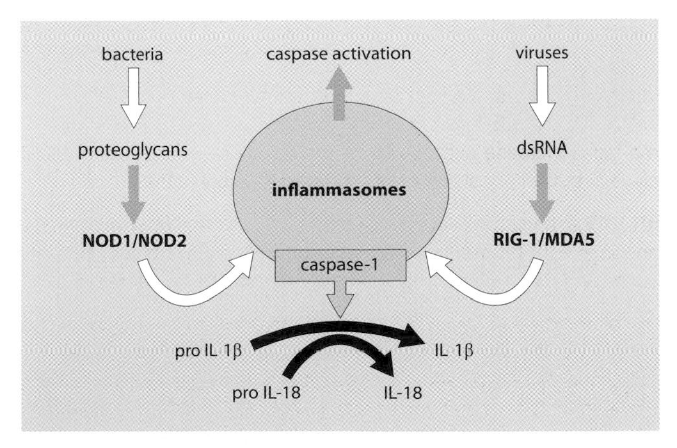

Fig. 3.23 Intracellular pattern recognition receptors.

CYTOTOXICITY

Cytotoxicity is a general term for the ways in which lymphocytes, mononuclear phagocytes, and granulocytes can kill target cells. This kind of interaction is important in the destruction of cells that have become infected with viruses or intracellular microorganisms, which they are unable to eliminate.

T cell-mediated cytotoxicity involves the recognition of antigenic peptides associated with MHC class I molecules (usually) on the surface of the target cell and is effected by $CD8^+$ Tc cells. The attacking cell orientates its granules toward the target and releases the contents, including perforin and granzymes, at the junction between the cells. Cytokines such as lymphotoxin, or the engagement of CD95 on the target, may also signal cell death. The relative contribution of each component depends on the cytotoxic cell involved. Target cell death occurs by apoptosis.

Fas (CD95) and **CD178 (CD95L)** Fas is a receptor belonging to the TNF-R family expressed on many cell types. Ligation of CD95 by CD95L (CD178) induces target cell death. Fas has an intracytoplasmic 'death' domain which occurs on other receptors involved in cell survival or death.

Perforin is a pore-forming molecule related to complement C9, which polymerizes on the target cell membrane to form channels.

Granzymes are serine proteases found in the granules of cytotoxic T cells, which may enter the target cell via perforin pores. Granzyme-A nicks DNA and prevents DNA repair, while granzyme-B activates caspases 3, 7, and 8, which induce apoptosis.

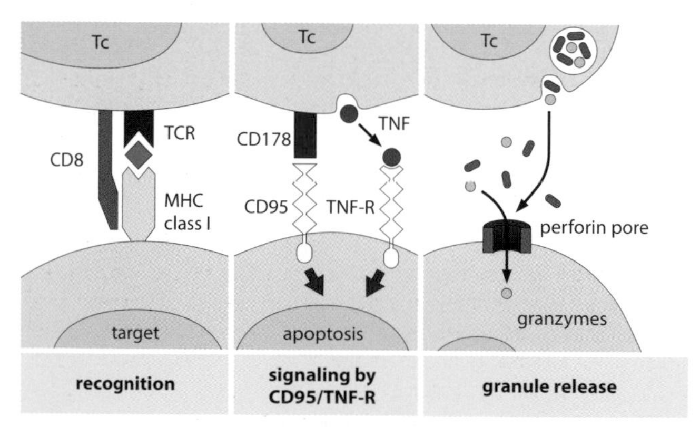

Fig. 3.24 Mechanisms of T cell-mediated cytotoxicity.

Caspases (cysteine aspartic acid proteases) are a group of proenzymes that become activated by cleavage into two or three subunits. They have a wide range of effects within the cell, affecting cell cycle control, DNA integrity and repair, and apoptosis. Ligation of Fas (CD95) by CD178, or the type I TNF receptor (TNFR-I, CD120a) by TNF-α or lymphotoxin, causes adaptor proteins to bind to the intracellular portion of the receptors and leads to activation of caspases 8 and 10. Activation of downstream effector caspases 3, 6, and 7 causes apoptosis.

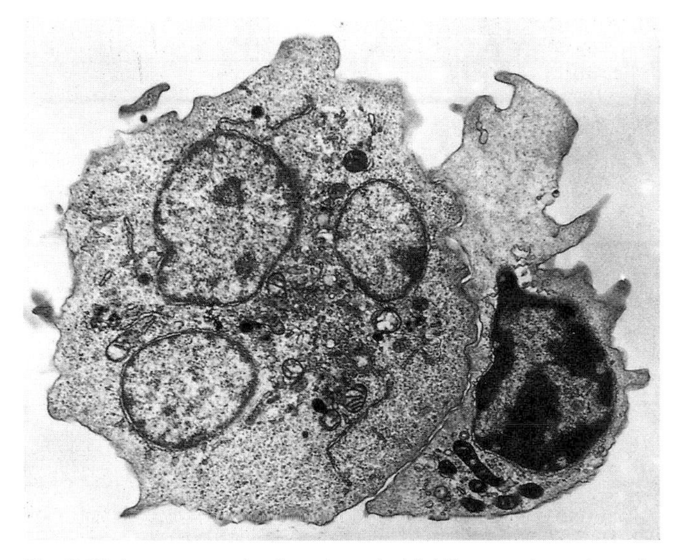

Fig. 3.25 Large granular lymphocyte (right) engaging a target cell (left). Courtesy of P. Penfold.

Antibody-dependent cell-mediated cytotoxicity (ADCC) involves the recognition of target cells coated with antibody. It may be effected by large granular lymphocytes, macrophages, or granulocytes, using their Fcγ receptors. The mechanism of cytotoxic damage depends on the effector cell; macrophages can release enzymes and ROIs, whereas LGLs use perforin and cytokines.

NK cell-mediated cytotoxicity is mediated by LGLs. They can kill some target cells that fail to express MHC class I or express allogeneic MHC class I. Thus they provide a line of defense against viruses that attempt to evade immune recognition by downregulating MHC expression. The mechanisms of cytotoxicity are similar to those used by Tc cells, with granule components (perforin and granzymes) being particularly important.

Eosinophil-mediated cytotoxicity Eosinophils are only weakly phagocytic and are less efficient than neutrophils and macrophages at destroying endocytosed pathogens. However, they can exocytose their granule contents, releasing factors that are very effective at damaging certain large parasites. Eosinophils recognize targets via bound antibody, including IgE, which they bind via FcεRII. Eosinophil degranulation is triggered by ligation of FcεRII or FcγRII. It is also induced *in vitro* by cytokines including IL-5, TNF, IFN-β, and PAF. Eosinophil granules include phosphatases, aryl sulfatase, and histaminase, in addition to those listed below.

Major basic protein (MBP) is a highly cationic protein that forms a major component of the crystalloid core of eosinophil granules. It is solubilized before secretion and can damage parasites. Figure 3.26 illustrates progressive damage to a schistosomule larva incubated in MBP. MBP also causes damage and loss of bronchial epithelium in allergic asthma.

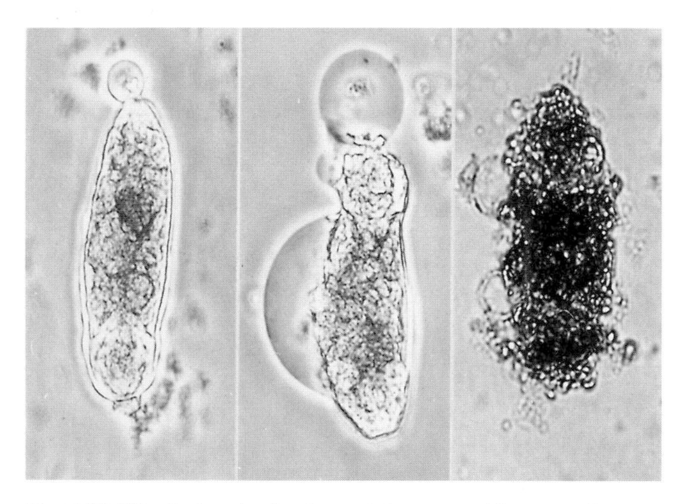

Fig. 3.26 Effect of major basic protein on a schistosomule larva. Courtesy of D. McLaren and Janice Taverne.

Eosinophil cationic protein (ECP) is a highly basic zinc-containing ribonuclease that binds avidly to negatively charged surfaces. It is particularly effective at damaging the tegument of schistosomes.

Eosinophil peroxidase is distinct from the myeloperoxidase produced by neutrophils and macrophages, but it serves a similar function in the generation of toxic hypohalites.

INFLAMMATION

Inflammation is the response of tissue to injury, with the function of bringing serum molecules and cells of the immune system to the site of damage. The reaction consists of three components:

- Increased blood supply to the region
- Increased capillary permeability
- Emigration of leukocytes from blood vessels into the tissues

Inflammation is an ordered process mediated by the appearance of intercellular adhesion molecules on endothelia and the release of various inflammatory mediators from tissue cells and leukocytes. Plasma enzyme systems are particularly important sources of inflammatory mediators. These include the complement, clotting, fibrinolytic (plasmin), and kinin systems. Also active are the mediators released by mast cells, basophils, and platelets, as well as the eicosanoids generated by many cells at sites of inflammation. Generally, neutrophils are the first cells to appear at acute inflammatory sites, followed by macrophages and lymphocytes, if there is an immunological challenge.

Vasodilation is the dilation of the local arterioles caused by the actions of mediators such as histamine on the smooth muscle in the vessel wall, allowing increased blood flow.

Transudate/exudate Normally, only small molecules pass freely through the capillary wall. The fluid that passes through is a transudate. If inflammation occurs, the endothelial cells are caused to retract, permitting larger molecules to pass out too. This fluid, which is also rich in cells, is an inflammatory exudate.

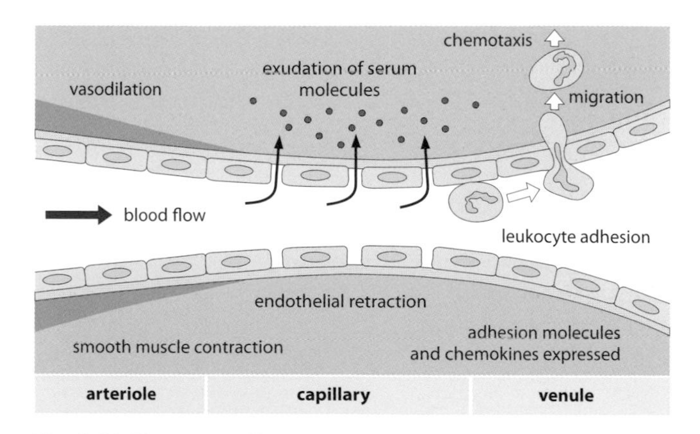

Fig. 3.27 Elements of inflammation.

mediator	origin	actions
histamine	mast cells basophils	increased vascular permeability smooth muscle contraction chemokinesis
5-hydroxytryptamine (5HT) = serotonin	platelets mast cells (rodents)	increased vascular permeability smooth muscle contraction
platelet-activating factor (PAF)	basophils neutrophils macrophages	mediator release from platelets increased vascular permeability smooth muscle contraction neutrophil activation
chemokines, e.g. CXCL8 (IL-8) CXCL10 (IP-10) CCL2 (MCP-1) CCL3 (MIP-1α) CCL5 (RANTES) CCL11 (eotaxin)	many cells, inc. endothelium mast cells leukocytes tissue cells	chemotactic for: neutrophils T cells, macrophages neutrophils, macrophages granulocytes, macrophages lymphocytes eosinophils
C3a	complement C3	mast-cell degranulation smooth muscle contraction
C5a	complement C5	mast-cell degranulation neutrophil and macrophage chemotaxis, neutrophil activation smooth muscle contraction increased capillary permeability
bradykinin	kinin system (kininogen)	vasodilation smooth muscle contraction increased vascular permeability pain
fibrinopeptides and fibrin breakdown products	clotting system	increased vascular permeability neutrophil and macrophage chemotaxis
prostaglandin E2 (PGE2)	cyclooxygenase pathway	vasodilation potentiates increased vascular permeability produced by histamine and bradykinin
leukotriene B4 (LTB4)	lipoxygenase pathway	neutrophil chemotaxis synergizes with PGE2 in increasing vascular permeability
leukotriene D4 (LTD4)	lipoxygenase pathway	smooth muscle contraction increased vascular permeability

Fig. 3.28 Mediators of acute inflammation.

Mediators of inflammation include the plasma enzyme systems, cells of the immune system, and products of pathogens themselves. The principal mediators are listed in Fig. 3.28.

Kinins are generated after tissue injury. Bradykinin is a nonapeptide produced by the action of kallikrein on high-molecular-weight kininogen. Lysyl bradykinin (kallidin) is generated by the action of tissue kallikrein on low-molecular-weight kininogen. The kinins are exceptionally powerful vasoactive mediators, causing vasodilation and increased capillary permeability.

Eicosanoids are mediators produced from arachidonic acid, which is released from membranes by the action of phospholipase A_2. Arachidonic acid is converted into eicosanoids by mast cells and macrophages, via two major pathways.

Prostaglandins (PG) and **Thromboxanes (Tx)** are produced by the action of cyclooxygenase on arachidonic acid. They have diverse proinflammatory effects, often synergizing with other mediators.

Leukotrienes (LT) are produced by the lipoxygenase pathway, which generates mediators of acute inflammation and factors important in the later phase of type I hypersensitivity.

Formyl-methionyl (f-Met) peptides (such as **fMLP)** are bacterial products that are highly chemotactic for neutrophils—bacteria initiate protein translation with f-Met, but eukaryotes do not.

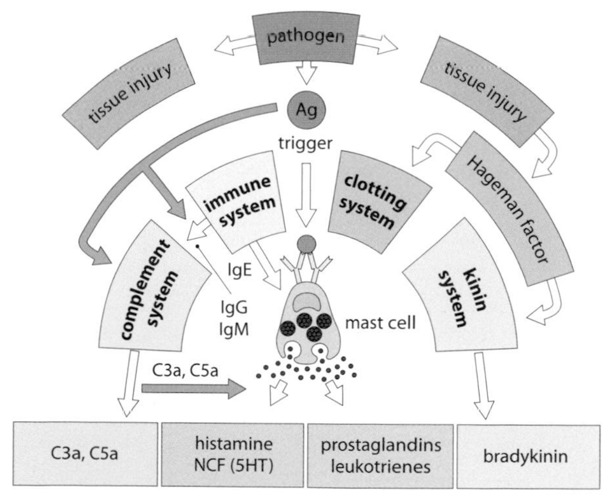

Fig. 3.29 Plasma enzyme systems.

MECHANISMS OF CELL MIGRATION

Leukocyte migration is controlled by molecules expressed on the surface of vascular endothelium that interact with complementary adhesion molecules on different populations of leukocytes. Most leukocyte migration occurs across venules. Several patterns of cell migration can be distinguished, including

- Movement of lymphocytes into secondary lymphoid tissues
- Migration of activated lymphocytes to sites of inflammation
- Migration of neutrophils into tissues during an acute immune response and migration of mononuclear cells into sites of chronic inflammation

Each pattern of migration is determined by particular sets of chemokines and adhesion molecules. There are three stages in the adhesion that precedes migration across the endothelium:

1. Slowing and rolling: most leukocyte migration occurs across venules as the shear force acting on circulating cells is lower and adhesion molecules are selectively expressed in venules. Initial slowing is mediated principally by selectins (such as E-selectin) on the endothelium, interacting with carbohydrate on the leukocytes.
2. Triggering: leukocytes that have been slowed may be triggered by chemokines released in the tissue or synthesized by the endothelium and bound to the endothelial cell surface. The chemokine signal is integrated over time, allowing cells to receive a sufficient signal to initiate migration. Triggering activates the integrins required for firm attachment to the endothelium.
3. Adhesion: the affinity of leukocyte integrins (such as LFA-1) on activated cells is increased, which allows them to bind to cell adhesion molecules (such as ICAM-1) induced on the endothelium by inflammatory cytokines. The integrins and CAMs are attached to the cytoskeleton of each cell, which allows the leukocyte to pull itself across the endothelium. Figure 3.30 shows a lymphocyte adhering to brain endothelium in encephalomyelitis.

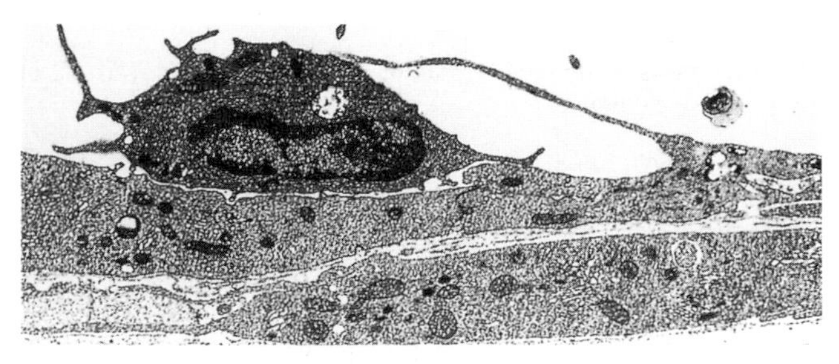

Fig. 3.30 Lymphocyte adhering to endothelium of the central nervous system. Courtesy of Clive Hawkins.

Diapedesis is the process by which adherent cells migrate across the endothelium and into tissues. Adherent cells extend pseudopodia into the junctions between endothelial cells, before squeezing through the gap. In tissues where endothelia have continuous tight junctions (for example, in the central nervous system (CNS)), migration occurs close to the junctions, but not through them. Enzymes released by the migrating cells dissolve the basal lamina. New adhesion molecules may now be mobilized to allow the cells to bind to cells of the tissue and extracellular matrix components.

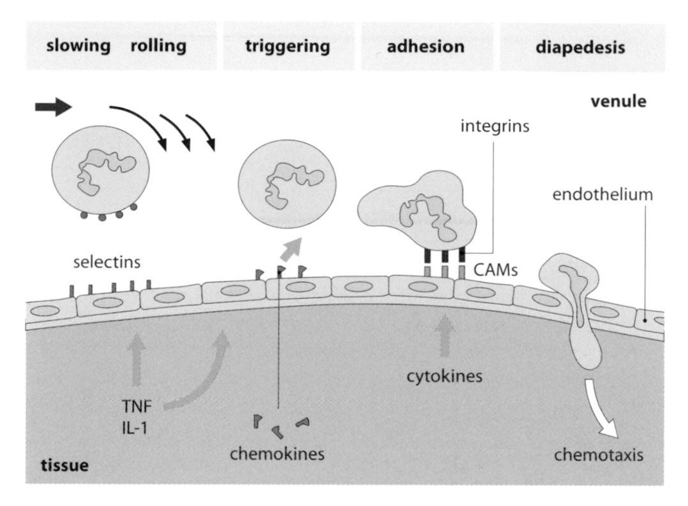

Fig. 3.31 Steps in leukocyte migration into tissues.

Chemotaxis is directional movement of cells in response to an inflammatory mediator. Cells are highly sensitive to, and migrate up, concentration gradients of molecules such as C5a, fMLP, and chemokines, if they have appropriate receptors.

Chemokinesis is increased random (nondirectional) movement of cells caused by inflammatory mediators, such as histamine.

Adhesion molecules belong to several different families. Some are constitutively expressed by cells (such as the integrin CR3 on mononuclear cells), whereas others may be induced by cytokines or cellular activation. Some adhesion molecules are retained in stores within the cells and may be quickly mobilized to the cell surface (such as LFA-1, stored in neutrophil 'adhesomes'); others (such as ICAM-1 in endothelium) must be synthesized. The major families of adhesion molecules are listed overleaf.

Selectins (CD62) are a group of three adhesion molecules with lectin domains, that can bind to carbohydrate. P-selectin and E-selectin, induced on endothelium, help to slow migrating leukocytes before adhesion. L-selectin is expressed on lymphocytes and neutrophils; on lymphocytes it contributes to their binding to high endothelial venules in mucosal tissues.

PECAM (CD31), expressed on endothelium, platelets, and some leukocytes, can undergo homotypic adhesion, which contributes to tissue integrity and may act as a guide during migration.

Integrins consist of an α and a β chain, both of which traverse the cell membrane. Usually, the α chain is unique to each molecule, but the β chain may be shared with other molecules. Adhesion is dependent on divalent cations; when Mg^{2+} is bound they adopt a high-affinity form. Integrins often have more than one ligand-binding site, recognizing different molecules. Several integrins bind target sequences, related to Arg-Gly-Asp (RGD), in the ligand molecule.

Leukocyte integrins are a family of three molecules that share the β_2 chain (CD18). They include LFA-1 (CD11a/CD18), important in migration of leukocytes across endothelium; CR3 (CD11b/CD18), expressed on all mononuclear phagocytes, which binds to ICAM-1 on endothelium at sites of inflammation; CR4 (CD11c/CD18), strongly expressed on tissue macrophages.

VLA (very late antigens) is the designation of the β_1 integrin family, which includes two molecules that appear late on activated T cells and may be involved in binding to extracellular matrix. VLA-4, which binds to VCAM-1, is used by lymphocytes migrating to sites of inflammation, particularly in skin and CNS.

CAMs (ICAM-1, ICAM-2, VCAM-1, and **MadCAM)** belong to the Ig supergene family. ICAM-1 and VCAM-1 are induced on endothelium by TNF, IL-1, and IFN-γ at sites of inflammation. ICAM-2 is constitutively expressed on endothelium and may control the base level of leukocyte traffic through a tissue. MadCAM-1, the mucosal addressin, binds to both L-selectin and integrins, to control migration into mucosal lymphoid tissues.

CD44 is a widely distributed adhesion molecule that can be produced in different splice variants, which determine its ligand-binding functions. During transendothelial migration it localizes to the leading pseudopod and can bind extracellular matrix.

molecule	structure	location	ligand(s)	function
P-selectin	selectin	endothelium neutrophils platelets	sLe^X = sialyl Lewis^X (carbohydrate)	acute inflammation neutrophil adhesion hemostasis
E-selectin	selectin	endothelium	sialyl Lewis^X (eg. CD15)	leukocyte slowing
L-selectin	selectin	lymphocytes neutrophils	sialyl Lewis^X	HEV binding slowing
ICAM-1	Ig family	endothelium (inducible)	LFA-1 CR3, CR4	adhesion and migration
ICAM-2	Ig family	endothelium	LFA-1	adhesion and migration
VCAM-1	Ig family	endothelium (inducible)	VLA-4 LPAM	adhesion
MAdCAM-1	Ig family sialylated	lymphoid endothelium	LPAM L-selectin	lymphocyte homing
PECAM	Ig family	endothelium lymphocytes	PECAM	adhesion activation migration guidance
LFA-1	$\alpha_L\beta_2$ integrin	leukocytes	ICAM-1 ICAM-2 CR3	migration
CR3	$\alpha_M\beta_2$ integrin	phagocytes	ICAM-1 ICAM-2 C3bi fibronectin	migration immune complex uptake
CR4	$\alpha_X\beta_2$ integrin	phagocytes	ICAM-1 ICAM-2 C3bi	adhesion immune complex uptake
VLA-4	$\alpha_4\beta_1$ integrin	lymphocytes	VCAM-1 LPAM fibronectin	adhesion at inflammatory sites and HEVs
LPAM	$\alpha_4\beta_7$ integrin	lymphocytes	MAdCAM-1	migration to lymphoid tissue
GlyCAM-1	sialoglycoprotein (soluble)	HEV	L-selectin	control of adhesion
PSGL-1	sialoglycoprotein	neutrophils	P-selectin	slowing in acute inflammation
CLA	glycoprotein	lymphocytes	E-selectin	lymphocyte migration to skin
VAP-1	sialoglycoprotein	HEV	L-selectin	lymphocyte homing
PNAd	sialoglycoprotein	HEV	L-selectin	lymphocyte homing

Fig. 3.32 Adhesion molecules for leukocyte migration.

CHEMOKINES AND CHEMOKINE RECEPTORS

Chemokines are a large group of cytokines that promote the chemotaxis and activation of a wide range of cells, including leukocytes. They are classified into four groups on the basis of their structure as α (CXC), β (CC), γ (C), and δ (CX3C)—the designation relates to the number and arrangement of conserved cysteine residues (C). Originally, they were given descriptive names such as macrophage chemotactic protein (MCP). However, these have been superseded by a system in which α chemokines are called CXCL1, CXCL2, etc., β chemokines are CCL1, CCL2, etc. For example, MCP-1 is now CCL2. Some chemokines are synthesized at sites of inflammation and control the migration of leukocytes across endothelium into inflamed tissues. Other chemokines are produced constitutively and control the normal movement of cells between lymphoid tissues and regions of these tissues, for example, between the cortex and germinal centers of lymph nodes. Figure 3.33 shows how chemokines can control the migration of different leukocytes into a site of inflammation. Inflammatory cytokines released in the tissue, such as TNF-α and IFN-γ, induce chemokine synthesis by the local endothelium, including CXCL8 (IL-8), acting on CXCR1, CCL2 (MCP-1) acting on CCR2, and CXCL10 (inflammatory protein-10, IP-10) acting on CXCR3. Which chemokines are produced depends on the tissue and the type of inflammation or immune response taking place. Chemokines can also be synthesized by tissue cells and transported to the endothelial surface. Each population of leukocytes has a different set of chemokine receptors, so the cells that enter a tissue differ, depending on the chemokines expressed on the endothelium.

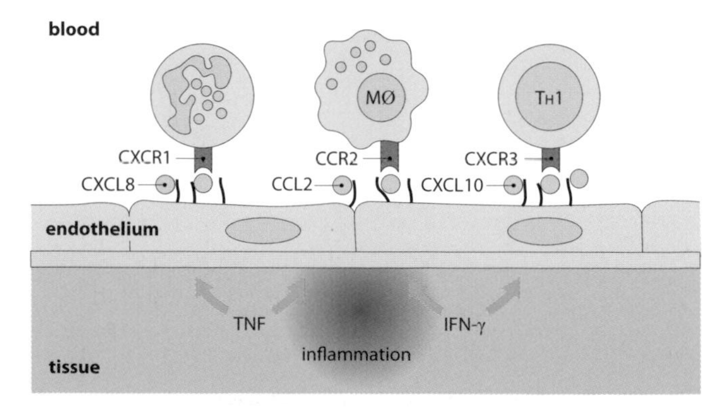

Fig. 3.33 Chemokine actions at sites of inflammation.

Chemokine receptors are designated according to which family of chemokines they recognize. For example α chemokines bind to CXCR1, CXCR2, etc. Most chemokines bind to several different chemokine receptors, and most receptors recognize several different chemokines. In addition, cells generally express several chemokine receptors so they can respond to a range of chemokines. Most cells of the body express some chemokine receptors at stages during their development, which control their position in the developing organism. Leukocytes change their receptors according to their state of differentiation and activation, which allows them to respond to inflammatory signals or position themselves in lymphoid tissues. For example, CCR7 is present on T cells, dendritic cells (DCs), and B cells. The chemokines that bind this receptor (CCL19 and CCL21) are produced in the T-cell areas of lymph nodes. Consequently T cells and DCs are attracted to these areas when they enter a lymph node. B cells can also be attracted to the T-cell areas when they express CCR7 after antigen stimulation. Figure 3.34 shows the complex pattern of chemokine receptor expression on leukocytes, but even this is simplified, because the relative expression is also important. For example, CXCR3 is found on T cells, but is highest on TH1 cells.

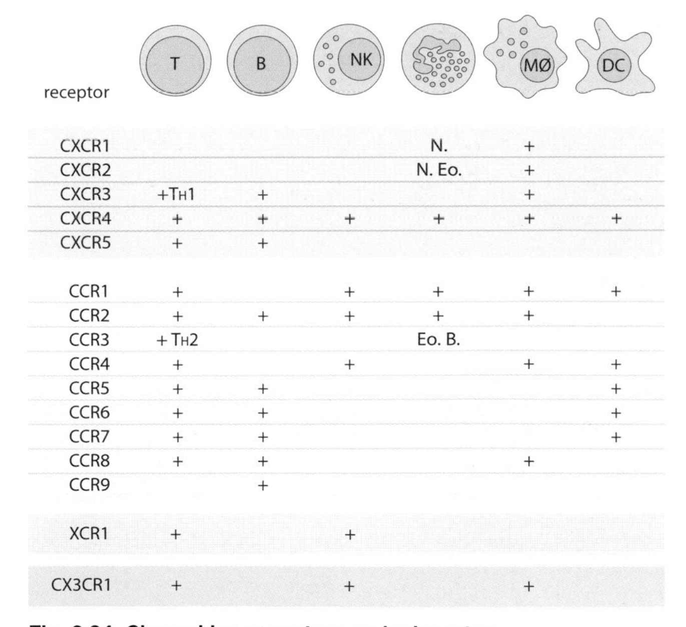

receptor	T	B	NK		MØ	DC
CXCR1				N.	+	
CXCR2				N. Eo.	+	
CXCR3	+TH1	+			+	
CXCR4	+	+	+	+	+	+
CXCR5	+	+				
CCR1	+		+	+	+	+
CCR2	+	+	+	+	+	
CCR3	+TH2			Eo. B.		
CCR4	+		+		+	+
CCR5	+	+				+
CCR6	+	+				+
CCR7	+	+				+
CCR8	+	+			+	
CCR9		+				
XCR1	+		+			
CX3CR1	+		+		+	

Fig. 3.34 Chemokine receptors on leukocytes.
N = neutrophil Eo = eosinophil b = basophil

COMPLEMENT

Complement is one of the serum enzyme systems. Its functions include mediating inflammation, opsonization of antigenic particles and microbes, and causing membrane damage to pathogens. The system consists of serum molecules, which may be activated via the classical, alternative, or lectin pathways. Molecules of the classical pathway are designated C1, C2, etc. Alternative-pathway molecules have letter designations, for example factor B (FB or just 'B'). The properties of the components are given overleaf, and their receptors on p. 74. The complement components interact with each other so that the products of one reaction form the enzyme for the next. Thus, a small initial stimulus can trigger a cascade of activity. Small fragments of complement molecules produced by cleavage are lower-case (C3a, C5b). Inactivated enzymes are prefixed 'i' (for example, iC3b) and active enzymes are indicated with a bar (for example, $\overline{\text{C3b,Bb}}$).

Classical pathway (yellow background) is activated by immune complexes binding to the C1q subcomponent of C1, which has six Fc-binding sites. C1q cleaves C1r and C1s. C1s then splits C4a from C4, and C2b from C2, leaving $\overline{\text{C4b,2a}}$, which can cleave C3.

Alternative pathway (Properdin pathway or **Amplification loop)** (purple background) is activated in the presence of suitable surfaces or molecules, including microbial products. C3b can bind either H or B. Normally H is bound and C3b is inactivated by I, but in the presence of activators B is bound and then enzymatically split by D, releasing Ba and leaving $\overline{\text{C3b,Bb}}$, which can cleave C3. This gives a feedback amplification loop to generate more C3b.

Lectin pathway (blue background) is activated by MBL or ficolins binding to bacterial carbohydrates.

C3 convertases, including $\overline{\text{C3b,Bb}}$ and $\overline{\text{C4b,2a}}$, clip C3a from C3 to leave C3b. C3b has a labile binding site that allows it to bind covalently to nearby molecules with –OH or $-NH_2$ groups. C3b together with a C3 convertase (such as $\overline{\text{C3b,Bb,3b}}$) can cleave C5.

Lytic pathway (orange background) is activated when C5b is deposited on membranes and associates with C6, C7, C8, and C9 to form the membrane-attack complex.

Membrane-attack complex (MAC) is a structure of C5b678 and polymeric C9, which traverses the target cell membrane and allows osmotic leakage from the cell.

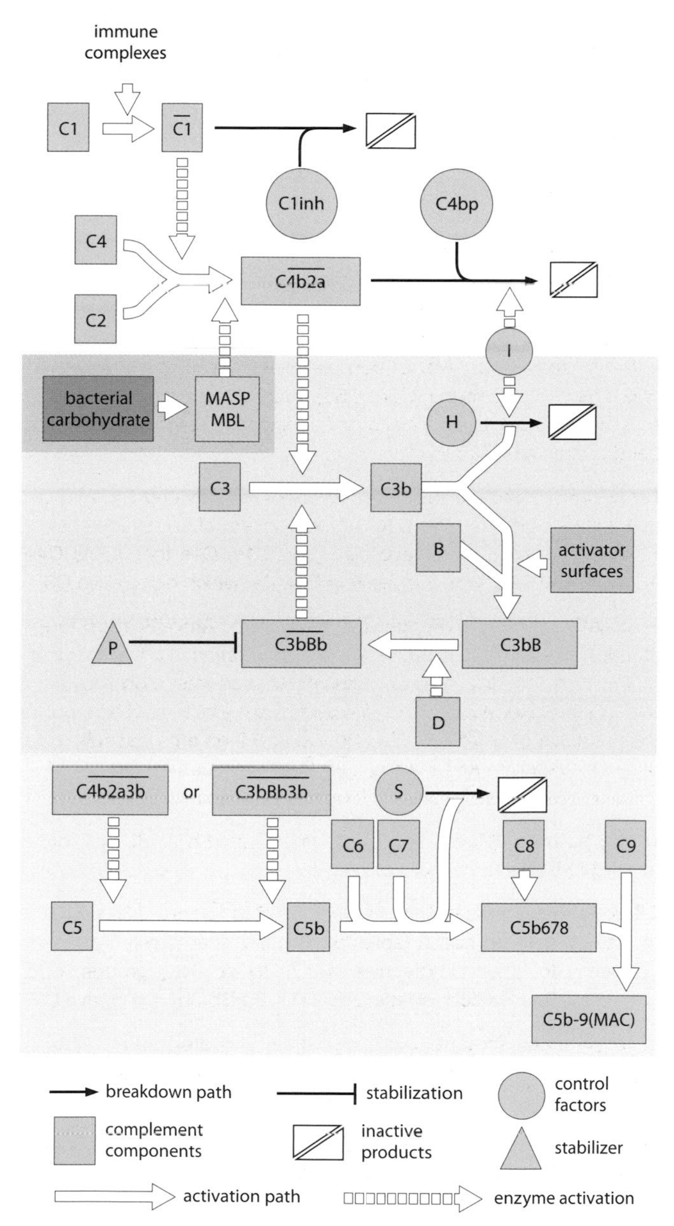

Fig. 3.35 Complement reaction pathways.

Complement fixation is the activation of complement, followed by deposition of the activated components on immune complexes or cell membranes. C3b and C4b can bind covalently to nearby molecules after cleavage of an internal thioester bond, which exposes a highly reactive group that can bind to –OH or $–NH_2$. The reactive group decays quickly by hydrolysis if a link is not formed. Hence complement is only deposited close to sites of activation.

Bystander lysis is the phenomenon whereby cells in close proximity to a site of complement activation have active components deposited on them and may then be lysed.

Anaphylatoxins C3a and **C5a** cleaved from the N-termini of the α chains of C3 and C5 mediate inflammation by causing mast-cell degranulation, smooth muscle contraction and increased capillary permeability. C5a is also chemotactic for neutrophils and monocytes. In this way, these peptides mimic some of the reactions of anaphylaxis. They are substantially inactivated by removal of their C-terminal arginine by serum carboxypeptidases.

Mannan-binding lectin (MBL) is a polymeric, pattern recognition molecule of the collectin family related to C1q. It binds bacterial and fungal carbohydrates and can activate the lectin pathway. MBL deficiency is associated in infants with respiratory infections.

Control of complement activation is effected by the natural decay of enzymatically active convertases and the actions of the various inhibitors and inactivators listed opposite. Membrane-associated molecules also alter the rate of complement breakdown; CR1 and DAF promote the decay of C3b,Bb.

Decay-accelerating factor (DAF, CD55) and **Membrane cofactor protein (MCP, CD46)** are proteins normally present on many mammalian cell membranes that limit the activity of the alternative pathway and the assembly of C5 convertases.

Protectin (CD59) is a membrane protein that protects host cells from lysis by binding to C5b678, to prevent polymerization of C9.

Paroxysmal nocturnal hemoglobinuria (PNH) is a condition in which red cell breakdown occurs via the alternative pathway. Patients' red cells are deficient in control proteins, particularly DAF.

Hereditary angioedema is due to a genetic deficiency of C1inh. There is uncontrolled local activation of C2, which undergoes conversion into a kinin that induces pathological local edema.

component	mol. wt (kDa)	serum conc. (μg/ml)	no. of poly-peptides	function
C1q C1r C1s	410 83 83	150 50 50	18 1 1	form a Ca^{2+}-linked complex—C1q $C1r_2$ $C1s_2$; C1q binds to complexed Ig to activate the classical pathway
C4 C2	210 115	550 25	3 1	classical-pathway molecules, activated by Cls to form a C3 convertase, C4b,2a
C3	180	1200	2	active C3 (C3b) opsonizes anything to which it binds and activates the lytic pathway. C3a causes mast cell degranulation and smooth muscle contraction. iC3b, C3d, C3e, and C3g are breakdown products of C3b
C5	180	70	2	C5b on membranes initiates the lytic pathway. C5a is chemotactic for macrophages and neutrophils, causes smooth muscle contraction, mast cell degranulation, and increased capillary permeability
C6 C7 C8 C9	130 120 155 75	60 50 55 60	1 1 3 1	lytic pathway components that assemble in the presence of C5b to form the membrane-attack complex and so may cause lysis
B D	95 25	200 10	1 1	B binds to C3b in the presence of alternative pathway activators, then is cleaved by D, an active serum enzyme, to form a C3 convertase C3b,Bb
P (properdin)	185	25	4	stabilizes C3b,Bb to potentiate amplification loop activity
MBL	540	1	18	binds bacterial carbohydrate and activates MASP-2
MASP-1 MASP-2	90 90	7 7	1 1	activates MASP-1 and MASP-2, and activates C4 and C2
C4bp H (β_1H) I (C3bina)	550 150 100	250 500 30	7 1 2	C4bp binds C4b, and H binds C3b to act as cofactors for I, which cleaves and inactivates C3b and C4b
C1inh	100	185	1	binds and inactivates $C1r_2$ and $C1s_2$
S-protein (vitronectin)	83	505	1	binds C5b-7 and prevents attachment to membranes

Fig. 3.36 The complement components.

IMMUNOREGULATION

The immune response is regulated primarily by antigen and co-stimulatory signals and secondarily by interactions between lymphocytes, APCs, and cells of the tissue. Antigen is the primary initiator of immune responses; the first signal required to trigger lymphocytes is antigen or antigen:MHC. Indeed, the immune system may be viewed as a homeostatic unit for the elimination of antigen. The essential role of antigen is seen at the cellular level. For example, antigen:MHC triggers T-cell activation and the expression of receptors for cytokines. Elimination of the antigen, by antibody or effector T cells, results in loss of the primary initiating stimulus, and the immune response is curtailed.

Danger signal is the idea that lymphocytes require both antigen stimulation and a 'danger signal' (co-stimulation) to become activated. The requirement for a dual signal acts as a fail-safe, to prevent unwanted immune reactions such as autoimmunity. In practice, danger signals are transduced by pattern recognition receptors (such as TLRs) that recognize microbial molecules.

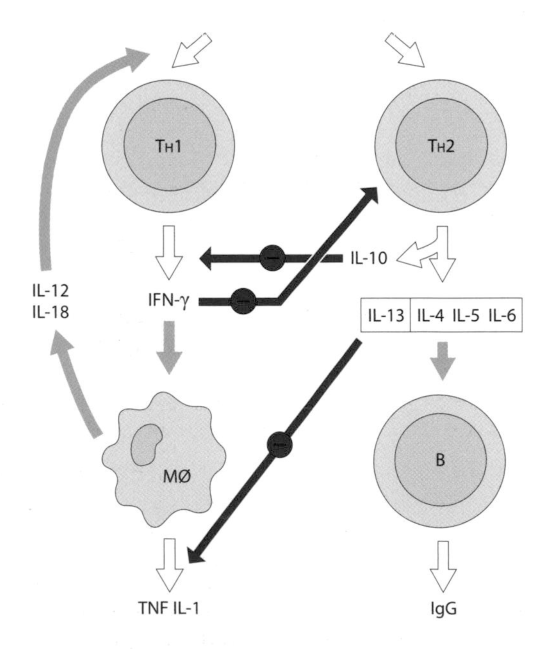

Fig. 3.37 Immunoregulation of TH1- and TH2-type responses.

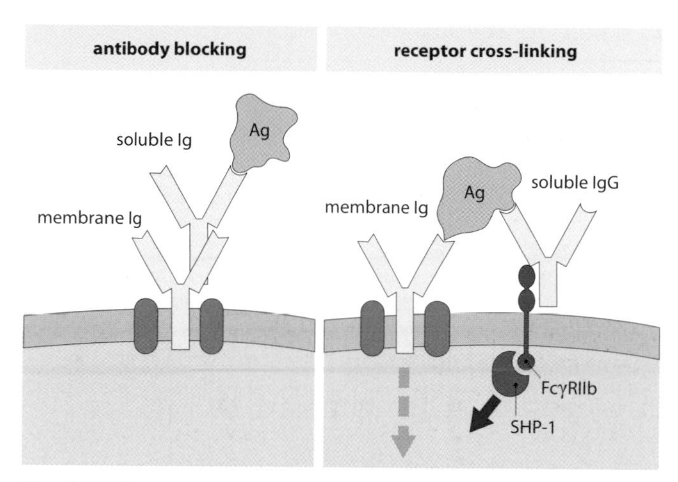

Fig. 3.38 Regulation of antibody production by antibody.

Antibody-mediated immunoregulation Antibody regulates its own production in several ways. Typically, IgM antibodies enhance the production of specific antibody, whereas specific IgG antibodies suppress further synthesis. The mechanisms include (1) binding to antigen, so preventing it from activating lymphocytes (Fig. 3.38, left); (2) binding to Fc receptors (FcγRIIb) on B cells, which in the presence of antigen cross-links the Fc receptors and surface Ig, delivering an inhibitory signal to the cells, mediated by the phosphatase SHP-1 (Fig. 3.38, right); (3) promoting immune complex formation and localization of antigen in germinal centers, so promoting Ig class switching and induction of B-cell memory.

Immune complex-mediated immunoregulation Complexes containing IgG generally suppress B-cell activation by the mechanisms shown in Fig. 3.38, whereas complexes containing IgM enhance it. In effect, IgM-containing complexes, produced early in an immune response, enhance the response, whereas IgG-containing complexes, produced after class switching, suppress it.

TH1- and TH2-type responses The TH1 subset promotes cell-mediated immunity, whereas TH2 cells promote antibody production, including IgE (Fig. 3.37). Moreover, each mode of response suppresses the other. IFN-γ produced by TH1 cells limits proliferation of TH2 and TH17 cells, whereas IL-12 and IL-18 from mononuclear phagocytes promote TH1 development. Conversely, IL-10 from TH2 cells prevents cytokine production by TH1 cells, and IL-13 inhibits cytokine production by macrophages.

Regulation/Suppression A group of functionally defined, regulatory T cells (TREG cells) control the activity of other lymphocytes. (Originally, a population of CD8⁺ T suppressor cells was thought to mediate this activity, but most TREG cells are CD4⁺.) TREG cells can develop naturally in the thymus or may be induced in the periphery during immune responses (induced TREG cells). They constitute 5–10% of peripheral T cells. Regulation is an active process and can be distinguished from tolerance by transferring suppression with T cells. Animals lacking TREG cells are susceptible to aggressive inflammation and autoimmunity in the gut and endocrine organs. The cellular basis for TREG action involves some or all of the following mechanisms: (1) reducing the co-stimulatory activity of dendritic cells, (2) release of anti-inflammatory cytokines (IL-10, TGF-β, IL-35), (3) modulating the mode of immune response (see Fig. 3.37), (4) consumption of IL-2, (5) a direct cytotoxic action on helper T and cytotoxic T cells.

Tissue-dependent regulation Immune reactions in tissues are controlled by regulatory cytokines (IL-10, TGF-β, etc.), eicosanoids and direct cell–cell interactions. Regulatory molecules include:

> **CD47**, a widely distributed molecule that interacts with signal-inhibitory regulatory protein-α (SIRPα); it recruits a phosphatase SHP-2 to the membrane, which inhibits lymphocyte activation.
>
> **Fractalkine (CX3CL1)**, a chemokine that can be produced in a membrane or secreted form, acting on the receptor CX3CR1. The soluble form is chemotactic, but the membrane form, present on neurons, contributes to suppression of microglia in the CNS.
>
> **CD200**, a 2 domain member of the Ig supergene family, expressed on keratinocytes and Langerhans cells. It binds to a receptor, CD200R1, found on myeloid cells and inhibits activation.

Network hypothesis is a theory that lymphocytes may be regulated by recognition of idiotypes on the antigen receptors of other cells or by idiotype-bearing antibodies. Such regulation is secondary to that mediated by antigen and cytokines, because of the redundancy in the immune system; if clones of lymphocytes are suppressed, their functions can be taken by other clones.

Psychoimmunology is a branch of immunology concerned with the interactions of the nervous, endocrine, and immune systems.

Sickness behavior describes the behavioral changes that occur in a person suffering from infection, including loss of appetite, reduced mobility, and extended sleep. Many of these changes have been related to the actions of IL-1 on the brain. IL-1 acts on temperature regulation centers in the hypothalamus to induce fever. It also suppresses appetite and induces slow-wave sleep.

Neuroendocrine regulation of immune responses seems to be important in damping immune and inflammatory responses, particularly via the production of corticosteroids.

Innervation of lymphoid tissues Thymus, spleen, and lymph nodes all receive sympathetic noradrenergic innervation, which controls blood flow through the lymphoid tissues, thus affecting lymphocyte traffic. However, fibers also run between the lymphocytes and seem to form junctions with individual cells. Denervation of lymphoid tissues can modulate immune responses.

Pituitary/adrenal axis Stress can induce release of adrenocorticotropic hormone (ACTH) from the pituitary. This induces the release of glucocorticoids, which are immunosuppressive. Lymphocytes also produce ACTH in response to corticotropin-releasing factor. In addition the adrenal medulla releases catecholamines, which can alter leukocyte migration patterns and lymphocyte responsiveness.

Endocrine and neuropeptide regulation Lymphocytes carry receptors for many hormones, including insulin, thyroxine, growth hormone, and somatostatin. These hormones, as well as encephalins and endorphins, released during stress, modulate T- and B-cell functions in complex, dose-dependent ways.

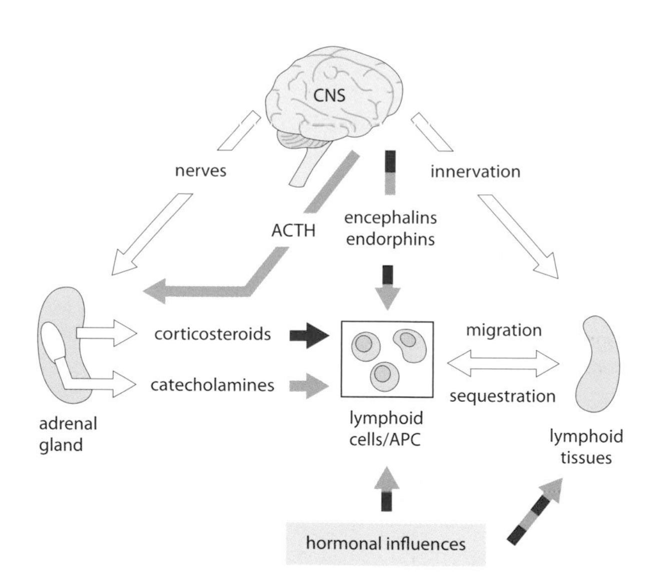

Fig. 3.39 Neuroendocrine regulation of immune responses.

TOLERANCE

Tolerance is the acquisition of nonresponsiveness to a molecule recognized by the immune system. Animals generally tolerate their own tissues: if they do not, autoimmune disease may result. Self-tolerance is thought to be due primarily to clonal deletion of cells in the neonatal period. As new mature lymphocytes develop they too are aborted, just when they are most susceptible to tolerization.

Neonatal tolerance Newborn animals are very susceptible to the induction of tolerance because of the general immaturity of their immune systems. Consequently, tolerance induced at this stage of life is very persistent.

Central tolerance refers to the induction of tolerance during lymphocyte development. Self-reactive T cells are deleted in the thymus, and self-reactive B cells in the bone marrow.

Peripheral tolerance is a necessary mechanism for maintaining tolerance to antigens that are not present in the primary lymphoid organs, or where the receptor is of low affinity.

B-cell tolerance In general, immature cells are more susceptible to tolerance induction than mature cells and can be tolerized by smaller doses of tolerogens. The dose of antigen and the way it is presented are critical. Self-reactive B cells fail to express Bcl-2 during development in the bone marrow or secondary lymphoid tissues and thus die by apoptosis. In the bone marrow, autoreactive B cells may escape deletion by editing their receptor specificity, done by making a new light-chain gene rearrangement. B cells may also become anergic to their antigen if they receive incomplete activation signals. Such cells downregulate surface IgM, while retaining IgD.

T-cell tolerance T cells are more easily tolerized than B cells. Once established, the duration of T-cell tolerance in an animal usually persists longer than for B cells. Immature T cells may be deleted during thymic development, although cells with low-avidity receptors remain. Mature T cells can be made anergic, depending on how antigen is presented to them (for example, lack of co-stimulation). Because B cells require help from T_H2 cells, B-cell tolerance may be a consequence of T-cell tolerance.

Superantigens are antigens that bind strongly to MHC molecules and can induce clonal deletion of T cells. Potentially they can modulate the T-cell repertoire.

High-zone and **Low-zone tolerance** Tolerance is best induced by high levels of antigen (high zone), which tolerizes B cells. However, some antigens in subimmunogenic doses (low zone) can also tolerize the T-cell population.

Mucosal tolerance and **Oral tolerance** Many antigens fail to induce an immune response when presented across the nasal mucosa as an aerosol, or across the gut mucosa in food (oral tolerance). The effect is dependent on the dose and frequency of the antigenic challenge. The effect may be due to the deviation of the immune response to TH2-type, with the production of suppressive cytokines and/or due to TREG activity.

Immune deviation refers to treatments aimed at switching the immune response from one mode to another (such as TH1 to TH2).

Tolerance mechanisms Several mechanisms maintain tolerance to self tissues (Fig. 3.40):

- Sequestration of antigen away from the immune system
- Central or peripheral tolerance induction of B and T cells
- Failure to process and present autoantigens by APCs
- Absence of co-stimulatory molecules on APCs
- Suppressive cytokines including IL-10 and TGF-β
- Direct and indirect actions of regulatory T cells

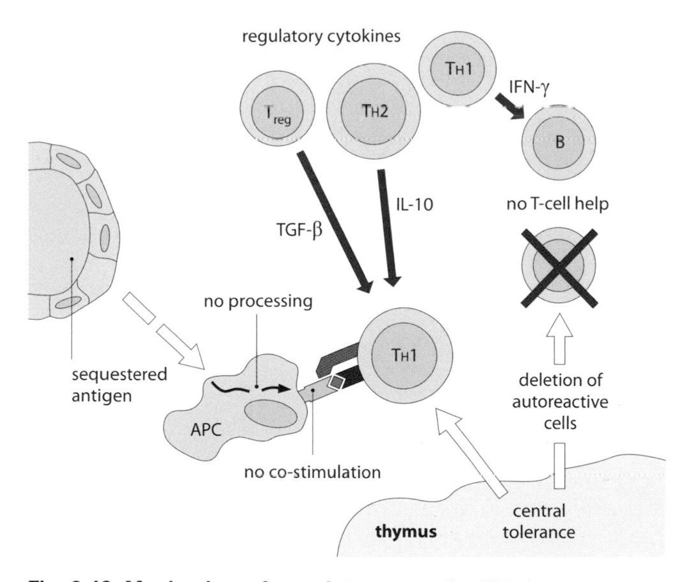

Fig. 3.40 Mechanisms for maintenance of self-tolerance.

GENETIC POLYMORPHISM IN THE IMMUNE RESPONSE

The genetic make-up of an individual affects their ability to mount an immune response. We can distinguish between mutations that clearly prevent the function of a particular component of the immune system and variants (polymorphisms) that affect the quality of an immune response. Polymorphisms are often stable in a population and many individuals have the variants, whereas mutations are deleterious and are usually lost over evolutionary timescales. The MHC is the most polymorphic of all gene loci, and variant MHC molecules differ in their ability to present antigen, leading to variations in immune response and disease susceptibility (Fig. 3.41). Polymorphism within the MHC is thought to reflect selection for variants that protect against particular infections, either extant pathogens or those that have occurred in historic times. One might therefore expect reports of particular MHC genes being associated with resistance to infectious disease. In practice there are many more reports of MHC haplotypes that are associated with susceptibility to infectious or autoimmune disease. This bias is seen because it is easier to identify positive disease associations and to get such results published. It is important to note that the relative risk associated with individual MHC genes (or haplotypes) depends on the population under investigation and the disease pattern. Susceptibility may be modified by other genes in the population, and in some cases the strain of pathogen can also affect whether an MHC gene is protective. Other polymorphisms affect the ability to respond to a wide range of antigens. For example, variants of TNF-α affect susceptibility to leprosy and severe cerebral malaria.

disease	resistant haplotype	population
HIV/AIDS	B53 DRB1*01 B44	USA, Hispanic Kenyan Chinese
hepatitis B	DRB1*1301 DR9 DRB1*1201	Gambian, German Korean Chinese
Plasmodium falciparum malaria	B35 B53	Malian Gambian
tuberculosis	DRB1*13 DRB1*11	Polish Chinese

Fig. 3.41 HLA polymorphism and disease resistance.

Immune response (Ir) genes were first identified in inbred strains of mice, according to their effect on the antibody response to defined antigens. The most important Ir genes encode MHC class II molecules, which determine antigen presentation to helper T cells. MHC class I molecules affect presentation to Tc cells and hence the ability to resist viral infections. There is also limited variation in MHC genes controlling antigen processing and presentation (DM, TAP, etc.). Genes encoding specific haplotypes of antigen receptors (*IGH* and *TCR*) have been linked to autoimmune conditions as well as restricting the responses to exogenous antigens (see clonal restriction below). Significant polymorphism is not confined to exons. For example, the promoter of the TNF-α gene is linked to autoimmunity in NZW mice. In addition, the promoters of MHC class II genes vary between strains, resulting in different responses to IFN-γ.

Repertoire is the sum total of antigen receptors produced by the immune system. The initial repertoire is partly determined by the genes of the TCR and antibody heavy and light chains.

Clonal restriction refers to an immune response produced by a limited number of clones. For example, the primary immune response to phosphocholine in Ig^a haplotype mice is dominated by the T15 idiotype. T-cell responses can also be clonally restricted, as a result of selective antigen presentation by particular MHC molecules expressed in a strain.

Biozzi mice are strains bred to give high or low antibody responses to an antigen (originally sheep erythrocytes). At least 10 non-MHC genes control responsiveness. The high and low responders differ in how their macrophages handle antigen; low responders degrade antigen quickly and do not present it well.

macrophage function	low responder	high responder
1. antigen uptake	+++	+
2. lysosomal enzyme activity	+++	+
3. intracellular degradation of antigen	+++	+
4. surface persistence of antigen	+	+++

Fig. 3.42 Macrophage functions in Biozzi mice.

IMMUNOSUPPRESSION

Immunosuppression describes measures used to reduce immune responses, particularly in transplantation surgery to prevent graft rejection, and in the control of autoimmune diseases. Most drug treatments are not antigen-specific, although some have greater effects on the immune system than other tissues.

Steroids including glucocorticosteroids, corticosteroids and synthetic steroids (such as dexamethasone) have numerous immune-suppressive and anti-inflammatory effects, macrophages being particularly sensitive. Steroids inhibit arachidonic acid release and hence reduce eicosanoid production. They also reduce secretion of neutral proteases and IL-1. Steroids interfere with antigen presentation, inhibit the primary antibody response and reduce the number of circulating T cells.

Azathioprine and 6-mercaptopurine are purine analogues that act on small lymphocytes and dividing cells, thereby blocking the development of effector cells. Monocytes are reduced in number and NK cell activity is also inhibited.

Cyclophosphamide and Chlorambucil are alkylating agents that damage DNA and prevent its replication. They act primarily on lymphocytes and strongly inhibit antibody responses, but have little effect on phagocytes. Experimentally, cyclophosphamide prevents B cells from regenerating their receptors.

Methotrexate is an analogue of folic acid that inhibits DNA synthesis and repair, and hence lymphocyte proliferation.

Mycophenolate inhibits the synthesis of guanosine. Lymphocytes are particularly susceptible to inhibition by this drug.

Cyclosporin-A is a fungal metabolite that interferes with cytokine production by T cells, particularly IL-2, and it inhibits IL-2R expression; both are early events in lymphocyte activation. It does not affect lymphoblasts, nor is it antimitotic. It is used to treat acute graft rejection, but has increasingly been replaced by less toxic drugs, listed below.

Tacrolimus (FK506) is a bacterial macrolide that prevents T-cell activation and IL-2 transcription by acting on calcineurin, an enzyme required for signal transduction from the T-cell receptor.

Rapamycin inhibits the ability of T-cell growth factors to put T cells into cell cycle. Rapamycin and tacrolimus bind to the same receptor, although their modes of action are different.

Antagonist peptides are analogues of peptides that bind to MHC molecules of particular haplotypes. They have been used as an experimental treatment for autoimmune conditions. By occupying the MHC-binding site, they block access for autoantigen peptides.

Antibody therapy is the use of monoclonal antibodies to treat disease, usually tumors or autoimmune diseases. Some are used to prevent graft rejection (Fig. 3.43). The first monoclonal antibodies were produced in mice, but these were potentially immunogenic in humans and less suitable for long-term therapy. Immunogenicity was reduced by genetic engineering to produce chimeric antibodies containing the antigen-binding V domains of the original antibody and human C domains. Alternatively, humanized antibodies, with antigen-binding hypervariable regions inserted into a human antibody gene framework, are even less immunogenic. Fully human antibodies can be produced in mice transgenic for human immunoglobulin genes.

antibody	type	target	treatments for
basiliximab daclizumab	chimeric humanized	IL-2R (CD25) IL-2R	graft rejection
belimumab	human	Blys (BAFF)	SLE
canikimumab	human	IL-1β	inflammatory disease
adalimumab certolizumab golimumab infliximab	human humanized human chimeric	TNF-α TNF-α TNF-α TNF-α	ulcerative colitis, Crohn's disease, RA ankylosing spondylitis autoimmune diseases
eculizumab	humanized	C5	paroxysmal nocturnal hemoglobinuria
muromonab-CD3	murine	CD3	graft rejection
natalizumab	humanized	VLA-4	multiple sclerosis, Crohn's disease
omalizumab	humanized	IgE	allergic asthma
tocilizumab	humanized	IL-6R	rheumatoid arthritis

Fig. 3.43 Monoclonal therapeutic antibodies.
SLE = systemic lupus erythematosus; RA = rheumatoid arthritis; BAFF = B-cell activating factor.

IMMUNOPOTENTIATION

Biological response modifiers (BRM) are compounds that modify an immune response, usually enhancing it. They include immunopotentiating bacterial and viral products (see p. 50 Toll-like receptors), and physiologically active molecules, including cytokines, as well as the true adjuvants, which are administered together with antigen. A number of these substances have been used in an attempt to potentiate immune reactions in cancer, by inducing cytokine production or the expression of co-stimulatory molecules on APCs. Bacterial products include:

BCG (Bacillus Calmette–Guérin), a live nonvirulent strain of *Mycobacterium bovis,* which is used in vaccines for immunization against tuberculosis.

Muramyl dipeptide (MDP), the smallest adjuvant active part of BCG, extractable from the cell wall.

Endotoxin/Lipopolysaccharide (LPS), a component of Gram-negative bacterial cell walls that is mitogenic for B cells and activates macrophages after binding to TLRs (see p.51, Fig. 2.22).

***Bordetella pertussis* toxin/toxoid (PTx)**, a lymphocytosis-promoting factor, which binds glycans on many cell types, particularly T cells, on which it acts as a mitogen.

Thymic hormones are factors produced by the thymus that assist T-cell development in the thymus and their maintenance in the periphery. They include thymosin, thymopoietin, thymostimulin, and thymulin (facteur thymique sérique).

Adjuvants are compounds that enhance the immune response, when administered with antigen, to produce higher antibody titers and prolonged production. The distinction between primary and secondary immune responses becomes blurred when adjuvants are used. Adjuvants typically consist of a depot of antigen, which may include bacterial components.

Freund's adjuvant is a water in oil emulsion containing the antigen (= Incomplete Freund's adjuvant). Complete Freund's adjuvant also includes dried, heat killed mycobacteria which induces very strong immune reactions and local necrosis. It is not used in any human vaccines.

Aluminum-adjuvanted vaccines use either aluminum hydroxide (alum) or aluminum phosphate mixed with the antigen, which becomes adsorbed on the surface of the gel. These adjuvants are used in many human vaccines, acting as antigen depots.

VACCINES

Vaccines are antigen preparations produced in a number of different ways, depending on the pathogen, its route of infection, and how it produces disease pathology. In addition to the antigen preparation and adjuvants (see opposite), vaccines often contain stabilizers and preservatives. Most vaccines are given by subcutaneous or intradermal injection, but some are given orally (such as rotavirus, polio (Sabin)) or nasally (for example, some flu vaccines).

Toxoids are chemically modified toxins, which retain antigenicity while destroying pathogenicity. They are used where the toxin produces the majority of the pathology (such as tetanus, diphtheria).

Attenuated live vaccines are live bacteria or viable viruses that have been modified to remove pathogenicity. They generally produce better immunity than killed organisms, but are more likely to produce adverse reactions. Also, because they may divide, they can be unsuitable for immunocompromised people.

Subunit vaccines consist of an antigenic subcomponent of the pathogen, produced either by fractionation or by biotechnology. For example, a subunit of hepatitis B isolated from blood was later superseded by the same antigen expressed in yeast.

Vector vaccines are produced by inserting genes for antigens of a pathogen into a nonpathogenic viral vector, such as vaccinia or adenovirus. Possible use in humans is under investigation.

Conjugate vaccines are used where the key antigenic component of the vaccine is only weakly immunogenic; for example, three polysaccharide antigens are coupled to diphtheria toxoid (carrier) in a conjugate vaccine for meningitis C. The toxoid component is presented to T cells, which help B cells make antibody against the polysaccharide antigens of the meningococcus.

DNA vaccines are experimental preparations, in which the DNA for an antigen, rather than the antigen itself, is used. The DNA is injected using a gene gun, and the technique relies on the DNA being taken up and expressed by cells of the recipient.

Combined vaccines Many vaccines are given in combinations, during early infancy. This is for convenience, to reduce the number of visits needed to an immunization clinic. Examples are the trivalent DPT vaccine—diphtheria, pertussis, tetanus. The UNICEF-recommended pentavalent vaccine also includes hepatitis B and *Haemophilus influenzae* B.

4 Immunopathology

IMMUNODEFICIENCY

Immunodeficiency is often identified in individuals by their increased susceptibility to infection, caused by a failure of one or more divisions of the immune system. Primary immunodeficiencies are inherited and may affect any part of the system. Examples include failure of lymphocyte development, impaired granulocyte functions, lack of macrophage receptors, and the absence of particular complement components. These deficiencies usually become apparent in the early months of life as immunity conferred by maternal antibodies wanes. Secondary or acquired immunodeficiency is a consequence of pathogenic infections, some of which directly attack the immune system (such as HIV), while others subvert immune responses (such as malaria).

Severe combined immunodeficiency (SCID) is a group of conditions with leukopenia, impaired cell-mediated immunity, low or absent antibody levels, and undeveloped secondary lymphoid tissues. About 25% of cases can be attributed to autosomal recessive adenosine deaminase deficiency or purine nucleoside phosphorylase deficiency. Around 50% of cases are due to lack

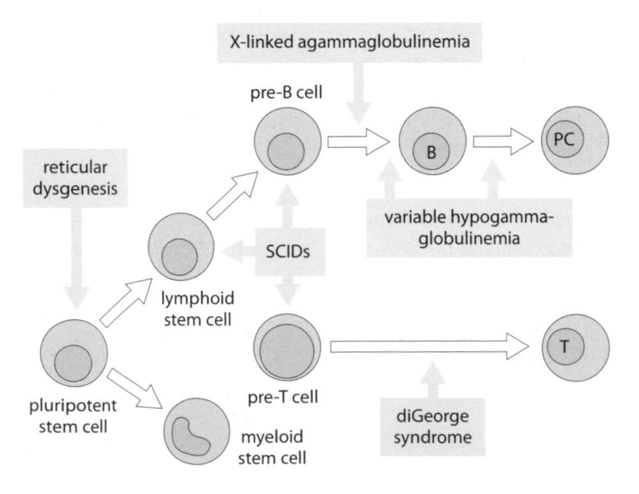

Fig. 4.1 Immunodeficiencies.

of the common γ chain for the receptors for IL-2, IL-4, IL-15, etc., which is encoded on the X-chromosome, and this accounts for the higher incidence of SCID in males than in females. The remaining causes of SCID are mostly rare autosomal recessive diseases, including mutations in the genes *Rag-1* and *Rag-2* (see p. 36), which are required to generate the antigen receptors on B and T cells. The different forms of SCID may be correlated with points on the lymphomyeloid differentiation pathways at which the genes would normally act (Fig. 4.1).

DiGeorge and **Nezelof syndromes**, caused by failed development of the third and fourth pharyngeal pouches, result in thymic hypoplasia, with low numbers of functionally active T cells although T cell numbers may rise to normal within 1–2 years. The syndromes are associated with distinctive facial characteristics, including wide-set eyes and a short philtrum.

Thymoma, a thymocyte neoplasia, is associated with immunodeficiency and a number of autoimmune diseases, including myasthenia gravis and hemolytic anemia.

MHC class II deficiency (Bare leukocyte syndrome) is caused by a lack of transcription factors that bind to the 5′ controlling regions of the MHC class II genes. The lack of MHC class II causes impaired T-cell education and antigen presentation. Patients have recurrent infections, particularly of the gastrointestinal tract.

Ataxia telangiectasia and **Nijmegen breakage syndrome** are rare recessive conditions affecting genes involved in DNA repair and rejoining (*ATM* and *NBS1*, respectively). Both conditions produce neurological disease and have a similar cellular phenotype, which produces immunodeficiency with reduced proportions of some Ig subclasses. Both genes seem to be required for the gene recombination that occurs when B cells switch their antibody class and chromosomal breaks occur in the Ig gene loci.

Wiskott–Aldrich syndrome (WAS) is a primary X-linked immunodeficiency caused by a mutation in the WAS protein (WASp) that causes defective actin polymerization. The deficiency affects the organization of the immunological synapse, with severely reduced T-cell responses to antigens. NK-cell activity and cell motility are also impaired. Lymphocyte numbers are near normal but antibody classes are abnormal; IgA and IgE are increased, IgG is normal and IgM is decreased; antibody is rapidly catabolized. Affected boys generally develop severe eczema and infections with pyogenic bacteria and opportunistic pathogens.

X-linked proliferative syndrome (XLP) results from a failure to control the actions of cytotoxic T cells after an infection with Epstein–Barr virus (EBV). Either the Tc cells fail to control the infection in B cells, allowing a fatal disease to occur, or the B cells are totally destroyed, resulting in agammaglobulinemia or lymphoid malignancies or aplastic anemia. The primary defect is in the gene SAP that acts as an adaptor for CD150.

X-linked agammaglobulinemia (Bruton's disease) Patients with this condition have normal T-cell function and cell-mediated immunity to viral infections, but have very low immunoglobulin levels and do not make antibody responses. The B cells fail to express a kinase, Bruton's tyrosine kinase (Btk), required for the maturation of pre-B cells into mature B cells.

X-linked hyper-IgM (HIGM) is due principally to a mutation in CD154, the ligand for CD40; this interaction is required for class switching. IgM is produced at high levels, but antibody responses do not mature and patients are susceptible to pyogenic infections and the development of autoimmunity. In about 30% of cases, immunodeficiency with high IgM is an autosomal recessive condition.

Common variable immunodeficiency (CVID) is a group of conditions with common symptoms but no single cause; defects of CD20 or CD81 are two possible causes. In other cases the defect seems to be due to inefficient interaction between T and B cells, with reduced levels of CD86 and CD25 on the B cell. CVID affects B-cell differentiation; B cells are present but do not develop into plasma cells, class switching is generally impaired, and somatic hypermutation is defective in a subset of patients. The number of memory B cells may also be reduced. Consequently, IgG, IgA, and IgE levels are low and IgM may also be reduced. Patients are susceptible to bacterial infections of the lungs and sinuses, and this usually becomes progressively more severe with age. Other organs may then be affected. Further corollaries include inflammatory conditions, autoimmune diseases, and lymphomas.

Leukocyte adhesion deficiency (Lad-1, Lad-2) is characterized by impaired neutrophil localization to tissues and impaired phagocytosis. Lad-1 is due to lack of CD18, the common β chain of the integrins LFA-1, CR3, and CR4, used in cell migration and phagocytosis. Lad-2 is due to defective glycosylation, resulting in lack of ligands for E- and P-selectin, needed for migration.

Chronic granulomatous disease (CGD) is due to a defect in NADPH oxidase, resulting in impaired oxygen-dependent killing by macrophages. Infection with pyogenic bacteria (particularly those producing catalase) occurs and macrophages accumulate at sites of chronic inflammation, forming granulomas.

Chediak–Higashi syndrome is a condition with impaired phagocyte responses to chemoattractants and reduced killing of phagocytosed bacteria. A cytoskeletal defect underlies the condition.

Acquired immune deficiency syndrome (AIDS) is caused by the retroviruses HIV-1 or HIV-2, which infect cells expressing CD4, including T cells and some APCs. The virus enters cells by first attaching to a chemokine receptor acting as a viral co-receptor. In the earliest phases of the disease, virus tends to infect mononuclear phagocytes by CCR5, but later viral variants develop that preferentially infect T cells. After infection, some individuals have a transient fever, which may develop into lymphadenopathy. Within weeks, specific antibodies can be detected in the blood (seroconversion) and virus levels in the blood decline. Over the course of months or years, the number of $CD4^+$ T cells gradually declines; once levels reach a critical threshold, opportunistic infections can develop as a result of the decline in T cell-mediated immunity. In the late stage of the disease, viral load increases in the blood as antibody declines.

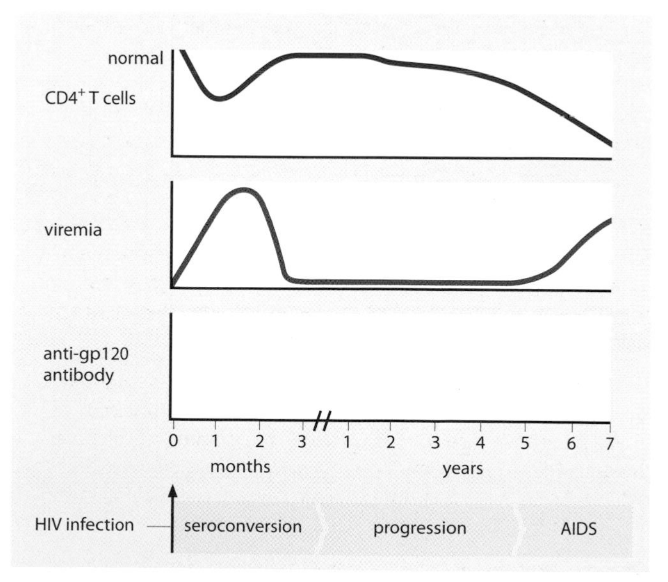

Fig. 4.2 Serology of HIV infection.

TRANSPLANTATION

Histocompatibility genes Grafts will usually be accepted if the recipient shares histocompatibility genes with the donor. A large number of gene loci affect graft rejection, but the MHC is the most important. Although it was first identified for its role in graft rejection, this is not its physiological function (see pp. 44–47).

Minor histocompatibility loci encode allelically variable proteins that induce weak graft rejection. Such molecules are processed and presented by the MHC class I molecules of the graft cells. In humans, even in MHC-matched transplants (such as between siblings), graft rejection reactions can still occur because of minor locus differences. Reactions induced by these antigens can usually be suppressed but MHC-dependent reactions are harder to control.

Passenger cells are donor leukocytes present in graft tissue. They are particularly important in sensitizing recipient TH cells to donor antigens because they express MHC class II molecules and can migrate into the host's lymphatic system.

First- and **Second-set rejection** The immune reactions that produce graft rejection display specificity and memory. For example, a skin allograft in humans will normally be rejected in 10–14 days, but if a second allograft of the same tissue type is given, the recipient will reject it faster, usually within 5–7 days (Fig. 4.3).

Rejection reactions are induced by recipient TH cells, which recognize allogeneic MHC molecules. These cells activate graft-infiltrating mononuclear cells to damage the graft. Alternatively, Tc cells can recognize allogeneic MHC class I and kill graft cells.

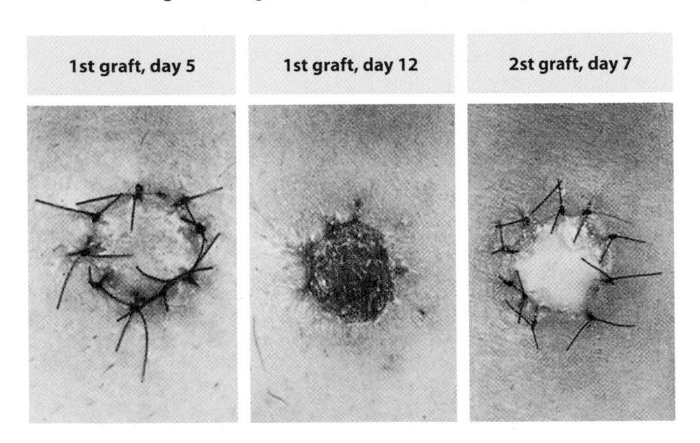

Fig. 4.3 Graft rejection displays immunological memory.

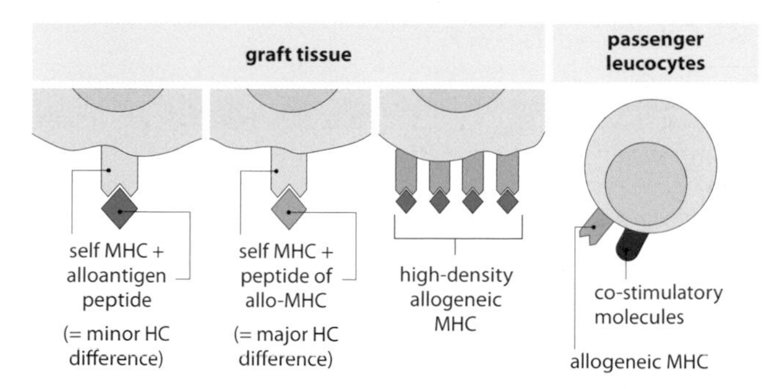

Fig. 4.4 Ways in which graft alloantigens may be presented.

Cross-matching To avoid graft rejection, the tissue type of the donor and recipient are cross-matched. All donor/recipient pairs are matched for the ABO blood group and for as many class I and II allotypes as possible. The greater the number that are shared (particularly class II), the higher the chance of graft survival.

Privileged tissues and sites Some allogeneic tissue grafts (such as liver) induce only weak immune reactions. In some cases this is due to low expression of MHC molecules. Privileged sites are areas where grafts are mostly isolated from the immune system, for example, the cornea of the eye lacks a lymphatic drainage.

Hyperacute/Acute/Chronic rejection describes the speed of rejection in organs such as the kidney. Hyperacute reactions occur within minutes of implantation and are caused by preformed antibodies against the graft. Acute rejection occurs within 2 weeks and is due to prior sensitization of the recipient to histocompatibility antigens. Chronic rejection occurs later and is due to the development of sensitivity to graft antigens. This sometimes occurs after the cessation of immunosuppression.

Graft-versus-host disease (GvHD) may occur when immunocompetent donor cells (for example from a bone marrow graft) recognize and react against the recipient's tissue, because the recipient is either immunosuppressed or cannot recognize the allogeneic cells. Sensitized donor T_H cells recruit macrophages to cause pathological damage, especially in skin, gut epithelium, and liver.

Enhancement includes ways of inducing tolerance in transplant surgery to enhance graft survival. The mechanism often involves interference with antigen presentation, for example by administration of anti-MHC class II antibody. Naturally occurring antibodies against the graft can also sometimes enhance its survival.

MHC DISEASE ASSOCIATIONS

Virtually every disease involving immune reactions is preferentially associated with particular haplotypes of MHC molecules. For example, individuals with the class I molecule HLA-B27 are 90 times more likely to develop ankylosing spondylitis than people lacking the allele. Figure 4.5 lists some of the conditions that show particularly strong MHC disease associations. It is important to understand that the haplotypes associated with a disease are specific for particular populations, and that a variant that confers susceptibility in one population may not do so in another; even if it does, the relative risk value will probably be different.

Shared epitope hypothesis can be used to explain the association of a disease with several different MHC haplotypes at the same locus, because they have a structural similarity. For example, DRB1*0401, *0404, *0405 and *0408 all share an epitope and are associated with increased risk of rheumatoid arthritis.

Relative risk (RR) is the risk of developing a disease when a particular HLA haplotype is present, compared with when it is absent. A relative risk more than 1 indicates that the haplotype is more prevalent in patients than the general population, whereas a relative risk less than 1 indicates that the variant may be protective. Note that RR is not the same as the 'odds ratio,' which is a different measure of the strength of association of two variables.

Linkage occurs between sets of genes on a single chromosome, such as the HLA complex. Unless crossover occurs between maternal and paternal chromosomes, a linked gene complex will be inherited as a block.

Linkage disequilibrium is the finding that some pairs of genes are found together more frequently than would be expected by chance, that is, more than the product of their gene frequencies. There are two possible explanations: (1) there is a selective advantage in inheriting the entire block of genes, or (2) two genes have appeared together by chance and there has been insufficient evolutionary time to separate them. Many sets of MHC molecules are linked, for example, HLA-A1 with HLA-B8, and HLA-A3 with HLA-B7. Consequently if one MHC molecule is associated with a disease, any linked haplotypes will also be associated with that disease, although they do not necessarily contribute to the disease susceptibility.

Extended haplotype refers to a set of linked genes, inherited as a block, for example HLA-DRB5*0101–DRB1*1501–DQA1*0102–DQB1*0602 is referred to as the haplotype DR2.

disease	haplotype	relative risk[†]
rheumatoid arthritis	DRB1*0401	6
	DRB1*0404	5
juvenile rheumatoid arthritis	DRB1*1402	47
	DRB1*1501	
ankylosing spondylitis	B27	87
Reiter's disease	B27	33
post-Shigella arthritis	B27	21
post-Salmonella arthritis	B27	18
Graves' disease	DRB3*0202	4
	B35	5
Hashimoto's thyroiditis	DQA1*0301	3
Addison's disease	DR3[§]	9
insulin-dependent diabetes	DQB1*0302	14
narcoplepsy	DQB1*0602	30
multiple sclerosis	DR2[§]	4
myasthenia gravis	B8	3
psoriasis vulgaris	B37	6
	B13	5
	DRB1*0701	9
Goodpasture's syndrome	DRB1*1501	13
chronic active hepatitis	B8	9
celiac disease	B8	8
	DQ2.5	20–60
dermatitis herpetiformis	B8	9
	DQ2	56
hemochromatosis	A3	8
	B14	5

Fig. 4.5 MHC disease associations (European Caucasians).
[†] Values vary between studies and populations.
§ Extended haplotypes.

MHC typing

MHC nomenclature MHC molecules are highly polymorphic, varying between individuals and loci. For example, the variant HLA-DRB1*0406 describes a variant of the HLA-DR gene at the B1 locus, which encodes the first of the DR-β chains. The variant produces molecules with serological specificity '04' and is the 06th genetic variant that produces this specificity. Additional figures can indicate expression (promoter) variants of the gene.

Tissue typing is the technique used to determine the MHC specificities carried by an individual. Typing is performed by adding antiserum of a defined specificity (such as anti-HLA-DR4) to the cell to be typed (usually lymphocytes). If the cells express the antigen, the addition of complement kills them, which can be detected by trypan blue staining.

Typing sera specific for particular MHC molecules were originally made by immunizing allogeneic individuals with cells and absorbing out unwanted specificities. This has been superseded by the use of haplotype-specific monoclonal antibodies.

Public (supratypic) and **Private specificities** If antigenic determinants are expressed on more than one MHC haplotype, it is a public specificity. Epitopes expressed on only one haplotype are private specificities.

Mixed lymphocyte culture/reaction (MLC/MLR) is a technique for typing cells, in which lymphocytes of different individuals are co-cultured. If the cells differ, they are stimulated to divide. The test can be done either with each set of cells reacting to the other (two-way MLR) or with one set (stimulator) treated so it cannot respond and only the proliferation of the responding (test) cells being measured (one-way MLR). A lack of response indicates that the test cell and the typing cell share an MHC specificity.

Homozygous typing cells, used in MLC, are cells with two identical MHC haplotypes. Human typing cells often come from the offspring of first-cousin marriages.

Primed lymphocyte typing test (PLT) is a highly sensitive MLC assay for detecting determinants that stimulate allogeneic T cells. The test cells are mixed with lymphocytes previously primed to a particular determinant by co-culture with homozygous typing cells. In a subsequent co-culture the primed cells proliferate rapidly if they encounter the priming MHC specificity again.

TRANSGENIC MICE

Transgenic strains are derived from a founder animal that has had new or variant genes inserted at the fertilized embryo stage. All cells of a transgenic animal carry the new genes. The tissue in which they are expressed can be determined by placing the transgene under the control of a tissue-specific promoter.

Knockout (null) strains are animals that have been genetically modified to remove particular genes involved in immune responses. They are particularly useful for determining the normal function of a gene, and many knockout strains model human immunodeficiencies. However, because the genes often function during development as well as in the immune system, the phenotype of knockout animals is sometimes unexpected. Figure 4.6 shows the effects of knocking out selected cytokine genes.

Knock-in strains are animals that have been genetically modified to overexpress particular genes, or to express them in all cells of a lineage. By using specific promoters the genes can be induced to express at a particular time with, for example, a drug activator.

cytokine	phenotype in mouse
TNF-α	defects in B-cell function abnormal lymphoid follicles very susceptible to intracellular pathogens
TNF-β (lymphotoxin)	no lymph nodes or Peyer's patches abnormal organization of spleen
IFN-γ	defective macrophage and NK activity reduced MHC expression very susceptible to intracellular pathogens
TGF-β	multiorgan inflammatory disease
IL-2	increased serum IgG increase in activated lymphocyte numbers inflammatory bowel disease
IL-4	low IgE and IgG1 deficient in cytokine production by T_H2 cells
IL-7	lack of peripheral lymphocytes and thymocytes
IL-12	impaired T_H1-type immune response reduced IFN-γ and increased IL-4 production

Fig. 4.6 Phenotypes of cytokine knockout mice.

ANIMAL MODELS AND MUTANT STRAINS

Before transgenic mice were developed, animal models of disease were identified either by noting strains that had a phenotype resembling the disease or by identifying sporadic mutants and inbreeding them to fix the mutation in the offspring. In general, mutants have been most useful for modeling single-gene immunodeficiencies, such as SCID. In contrast, strains have more often been used to model polygenic conditions, such as autoimmune conditions; susceptibility to autoimmune disease depends on many gene loci that interact with each other and environmental factors. Because of the complexity of such traits, it must be emphasized that the model may only resemble the human condition in its appearance. Figure 4.7 lists some of the more important model strains.

Inbred strains of animals are made by repeated brother–sister matings in successive generations, to give a strain with identical sets of autosomes. If by chance a pair of identical chromosomes occurs in the F_1 animals, inbreeding ensures that the pair remains fixed in the genome of subsequent generations. By repeated inbreeding, all of the chromosome pairs eventually become, and will remain, homozygous.

Recombinant strains are produced by crossing different inbred strains. On rare occasions, crossing-over occurs in the F_1 animal, so that the affected chromosome has different haplotypes at each end. These strains are used to identify the segment of chromosome responsible for a particular characteristic.

Recombinant inbred strains are produced by crossing strains (a × b) and then inbreeding the offspring. This gives strains that have identical sets of chromosomes, but each set will be of either the a-type or the b-type at random. They can be used to determine which chromosomes carry the genes for each trait.

Congenic strains are bred to be identical to each other except at some chosen locus. For example, an H-2^k congenic animal would have the MHC locus of the k haplotype superimposed on the background genes from a non-H-2^k strain.

Consomic strains are a special type of congenic strain that has an entire chromosome from one parental strain and the remaining chromosomes from the other. Such strains are useful for identifying the chromosomal origin of a particular trait.

strain/species	characteristics
nude mouse, nude rat	the nude mutants (*nu*) lack a thymus and all T cells a linked locus produces hairlessness
beige mouse (Bg)	NK cell and granulocyte defects affecting degranulation, elastase, and cathepsin G
NZB mouse	autoimmunity with hemolytic anemia and impaired immunoregulation (polygenic)
(NZB × NZW) F_1	autoimmunity with immune complex nephritis, used as SLE model (polygenic)
MRL.lpr or gld mouse	T-cell lymphoproliferation the lpr mutation affects CD95 (fas), and the gld mutation, CD95L (CD178)
Nod mouse (non-obese diabetic)	autoimmune reaction to pancreatic β cells model of type II diabetes (polygenic)
BXSB mouse	Y-chromosome linked Yaa mutation accelerates autoimmunity
SCID mouse	fails to recombine Ig or TCR genes due to defect in DNA repair enzyme, DNA-PKcs
CBA/N mouse	lacks CD5 B-cell subset X-linked deficiency (Xid) in a kinase (btk) required in Ig and CD 40 signaling
C3H/HeJ mouse	B cells lack receptor for LPS
DBA/2	impaired B-cell development mutation in kinase domain of cKit
motheaten (*me^{v}*) mouse	severe B-cell deficiency lacks a protein tyrosine phosphatase (PTP1c)
BB rat	spontaneous autoimmune diabetes and thyroid autoimmunity
buffalo rat	a proportion develop autoimmune thyroiditis and/or diabetes
obese chicken	autoimmune thyroiditis—model of Hashimoto's disease

Fig. 4.7 Characteristics of immunologically aberrant strains.

AUTOIMMUNE DISEASE

Autoantigens are self molecules recognized by T or B cells. B cells may be stimulated to make autoantibodies against them.

Autoreactive cells are lymphocytes with receptors for autoantigens. These cells can potentially produce an autoimmune response, but they do not necessarily do so.

Autoimmunity is the reaction of the immune system against the body's own tissues. To understand how autoimmune reactions can develop, it is necessary to know the mechanisms by which self-tolerance is normally maintained. These include:
(1) sequestration of autoantigen, (2) deletion of autoreactive lymphocytes in thymus and bone marrow, (3) failure to process and present some self molecules, (4) induction of anergy in T cells because of lack of co-stimulatory signals, (5) regulatory T cells, (6) suppressive cytokines and hormones (see Fig. 3.40, pp 100–101).

T-cell bypass Most self-reactive T cells are deleted or anergized, but autoreactive B cells may become activated by a mechanism that bypasses tolerant T cells. For example, a cross-reactive exogenous antigen taken up by an autoreactive B cell could be presented to a T cell recognizing a nonself epitope, which then helps the B cell (Fig. 4.8a). Alternatively, polyclonal activators such as Epstein–Barr virus could stimulate B cells directly.

T-cell autoreactivity may also be induced by cross-reacting microbial antigens. In Fig. 4.8b microbial adjuvants (such as LPS) induce co-stimulatory molecules on the macrophage, activating a quiescent autoreactive T cell. In Fig. 4.8c an enveloped virus is internalized by a macrophage and processed in the class II pathway. The viral envelope contains self molecules that are now presented. In Fig. 4.8d a quiescent self-reactive T cell is stimulated by a cross-reactive microbial antigen. After priming, the T cell expresses co-stimulatory molecules and is more readily activated if presented with self antigen.

Autoregulatory failure A breakdown in central or peripheral tolerance may also produce autoimmunity (Fig. 4.8e).

Organ-specific autoimmune diseases are directed primarily at particular tissues, for example, anti-thyroglobulin in Hashimoto's thyroiditis, or antibodies against pancreatic β cells in diabetes. Organ-specific autoantibodies and disease tend to occur together in individuals and they cluster in families, as a result of genetic predisposition.

Organ nonspecific autoimmune diseases are directed to widely distributed autoantigens, such as anti-DNA antibody in systemic lupus erythematosus. These conditions often produce type III, immune complex-mediated, hypersensitivity reactions.

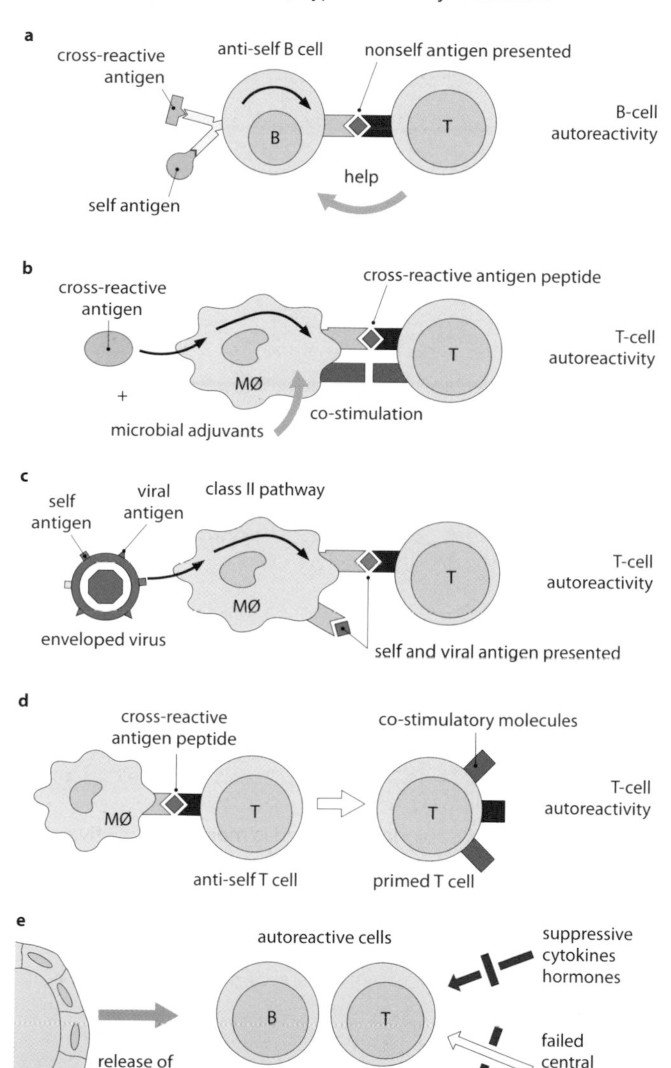

Fig. 4.8 Mechanisms for the breakdown of self-tolerance.

HYPERSENSITIVITY

Hypersensitivity is an immune response that occurs in an exaggerated or inappropriate form. In some cases responses may occur against innocuous external antigens, such as pollen in hayfever. In other cases responses against genuine pathogens are out of proportion to the damage caused by the pathogen. Also of great importance are the different kinds of tissue damage seen in autoimmune diseases: these are in effect hypersensitivity reactions, because any response to a self molecule is inappropriate. The hypersensitivity reactions were classified by Gell and Coombs according to the speed of the reaction and the immune mechanisms involved. Although they are classified separately, they do not necessarily occur in isolation and several mechanistically different reactions may be included in one type.

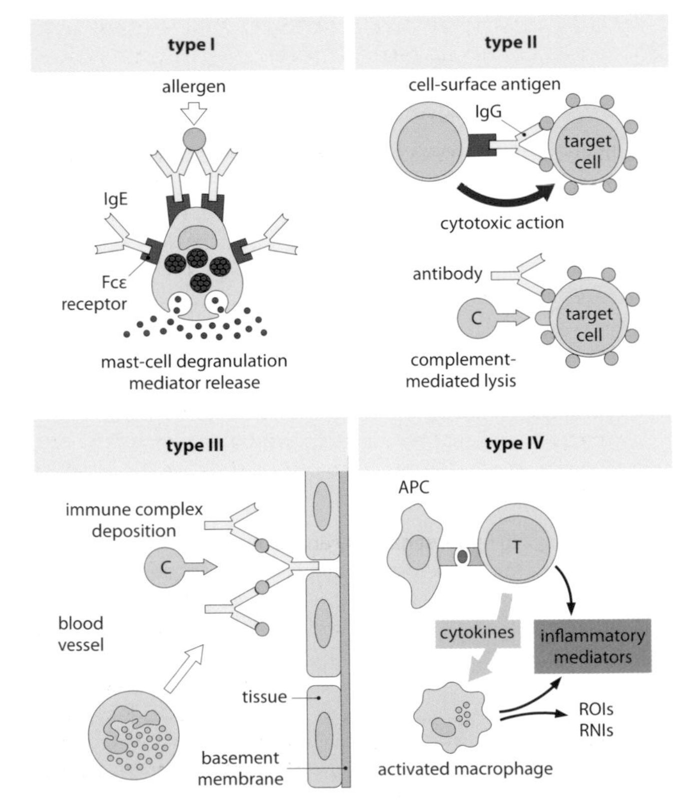

Fig. 4.9 Four types of hypersensitivity reaction.

Type I (Immediate) hypersensitivity is seen in allergic asthma, hayfever, and some types of eczema. It develops within minutes of exposure to antigen and is dependent on the activation of mast cells and the release of mediators of acute inflammation. Mast cells bind IgE via their surface FcεRI receptors; when antigen cross-links the IgE, the mast cells degranulate, releasing vasoactive amines and chemokines (see p.84, Fig. 3.28). Prostaglandins and leukotrienes, produced by arachidonic acid metabolism, contribute to a delayed component of the reaction that often develops hours after the original exposure to antigen.

Type II (Antibody-mediated) hypersensitivity is caused by antibody against cell-surface antigens and components of the extracellular matrix. These antibodies can sensitize the cells for antibody-dependent cytotoxicity (by macrophages or NK cells) or for complement-mediated lysis. Type II hypersensitivity is seen in the destruction of red cells in hemolytic disease of the newborn and in autoimmune hemolytic anemia. Tissue destruction in autoimmune diseases such as myasthenia gravis, Goodpasture's syndrome, and pemphigus is primarily antibody mediated.

Type III (Immune-complex-mediated) hypersensitivity is caused by the deposition of antigen:antibody complexes in tissue and blood vessels. This tends to occur at sites of filtration such as the kidney glomerulus and ciliary body of the eye. The complexes activate complement and attract polymorphs and macrophages to the site. These cells may exocytose their granule contents and release reactive oxygen and nitrogen intermediates to cause local tissue damage. The antigens in the complexes may come from persistent pathogenic infections (such as malaria), from inhaled antigens (for example, in extrinsic allergic alveolitis) or from the host's own tissue in autoimmune disease. These conditions are all characterized by a high antigen load, which may be associated with a weak or ineffective immune response.

Type IV (Delayed) hypersensitivity arises more than 24 hours after encounter with the antigen and is mediated by antigen-sensitized $CD4^+$ T cells, which release cytokines, attracting macrophages to the site and activating them. The macrophages produce damage, which may develop into chronic granulomatous reactions if the antigen persists. This type of hypersensitivity is seen in skin contact reactions and the response to some chronic pathogens, such as *Mycobacterium leprae*, *M. tuberculosis* and some *Schistosoma* species. Regulatory T cells would normally contribute to the control of such T cell-mediated reactions.

TYPE I (IMMEDIATE) HYPERSENSITIVITY

Allergy, originally meaning altered reactivity on a second contact with an antigen, now means type I hypersensitivity. The reactions are mediated by IgE and indicate a TH2-type response.

Allergens are antigens that induce type I reactions. Typical allergens are pollens, house dust mite feces, fungal spores, and animal skin flakes, which are 3–30 μm in diameter. They are inhaled in small quantities and are deposited on the mucous membranes of the nasal passages and airways.

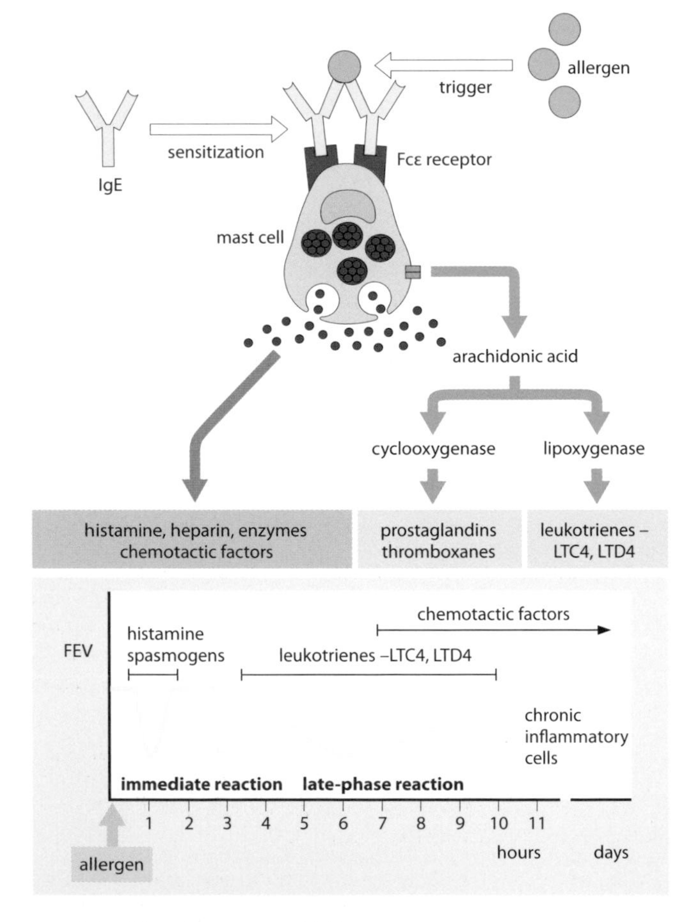

Fig. 4.10 Type I hypersensitivity.

Sensitization in this context is the process by which a susceptible individual develops an allergen-specific IgE response. The IgE binds to high-affinity IgE receptors on mast cells, thereby sensitizing them for triggering by antigen.

Triggering of mast cells occurs when an antigen cross-links the cell-surface IgE, causing an influx of Ca^{2+}, resulting in degranulation and activation of phospholipase A_2. Mast cells can also be directly triggered by anaphylatoxins (C3a and C5a), and some drugs (such as opiates, vancomycin).

Phospholipase A_2 is a membrane-associated enzyme that releases arachidonic acid, the substrate for lipoxygenase, which leads to production of leukotrienes and cyclooxygenase, and thence to synthesis of prostaglandins and thromboxanes.

Atopy describes conditions that manifest type I hypersensitivity, including asthma, hayfever, and eczema. They tend to cluster in families. Gene loci that are associated with increased risk of atopy include HLA, the cytokines IL-4, IL-5, IL-10, and Il-13, and receptors for leukotrienes (LTRI, LTRII) and chemokines (CCR3).

Immediate and late-phase reactions After bronchial provocation with an allergen, there is an immediate reduction in airway patency, measured as a fall in forced expiratory volume (FEV), caused by histamine, prostaglandins, and kinins and via the action of PAF on platelets. After several hours a late-phase reaction develops, caused primarily by leukotrienes and chemokines. Inflammatory cells including macrophages, basophils, and other polymorphs are attracted to the site. Eosinophil granule proteins are highly toxic for airway epithelium. Analogous immediate and late reactions occur in allergic skin reactions.

Anaphylaxis is a systemic type I reaction seen in sensitized animals injected with allergen. The release of vasoactive amines and spasmogens causes smooth muscle contraction, increased vascular permeability, and a fall in blood pressure. Respiratory or circulatory failure may ensue. Anaphylactic reactions may occur in humans, for example caused by bee venom or an adverse reaction to a vaccine component in a sensitive individual.

Desensitization is a treatment aimed at reducing levels of allergen-specific IgE by giving graded doses of the allergen over several months. The protocol deviates the immune response away from the TH2 type and induces higher levels of IgG.

Prick test is used to determine an individual's (type I) sensitivity to allergens, which are pricked onto the skin. Sensitive individuals develop a wheal-and-flare reaction.

TYPE II (ANTIBODY-MEDIATED) HYPERSENSITIVITY

Type II hypersensitivity is caused by antibody directed against membranes and cell-surface antigens. Complement may be activated and effector cells with Fcγ and C3 receptors can then engage the target tissue. Membrane-attack complexes may also be formed, to potentiate the damage. The site of damage depends on the antibodies involved.

Transfusion reactions occur when mismatched donor blood is infused into a recipient. The recipient may have naturally occurring antibodies against the foreign cells, as happens with the ABO blood group system, or these may develop after transfusion of allogeneic cells. The antibodies can cause complement-dependent lysis or sequestration of the sensitized cells in spleen and liver.

Blood groups are systems of allotypically variable erythrocyte surface antigens, some of which also occur on other tissues. The more common ones are listed in Fig. 4.11.

system	gene loci	antigens	phenotype frequencies
ABO	1	A, B, or O	A 42% B 8% AB 3% O 47%
Rhesus	3 closely linked loci: major antigen = RhD	C or c D or d E or e	RhD^+ 85% RhD^- 15%
Kell	1	K or k	K 9% k 91%
Duffy	1	Fy^a, Fy^b, or Fy	Fy^aFy^b 46% Fy^a 20% Fy^b 34% Fy 0.1%
MN	1	M or N	MM 28% MN 50% NN 22%
Lutheran	1	Lua or Lub 18 antigens	Lua <1% Lub >99%

Fig. 4.11 Six major blood group systems.

Hemolytic disease of the newborn (HDNB) is caused by maternal IgG antibodies against fetal red cells, which cross the placenta and destroy them. The mother becomes sensitized by fetal red cells entering her circulation at birth, so the first child is usually unaffected. The most common cases involve Rhesus-negative mothers carrying Rhesus-positive children, but the incidence of HDNB due to other groups (such as Kell) is significant. The risk of HDNB is reduced if the fetus also has a different ABO blood group. This observation underlies rhesus prophylaxis.

Rhesus prophylaxis is the administration of anti-Rhesus D antibodies to Rhesus-negative mothers immediately after they have delivered a Rhesus-positive child, so as to destroy the Rh-positive cells and prevent them from sensitizing the mother.

Autoimmune hemolytic anemia is caused by autoantibodies against red cells that cause their destruction. The antibodies may be either 'warm agglutinins,' which cause red cells to be removed by sequestration, or 'cold agglutinins,' which cause complement-dependent lysis. The antibodies are described according to the temperature at which they bind. Cold-reactive antibodies are usually specific for the Ii blood group system and cause red cell destruction in the peripheral circulation, particularly in winter.

Drug-induced reactions can occur when a drug or immune complexes containing a drug adsorb on red cells or platelets and induce complement-dependent lysis, resulting in anemia or thrombocytopenia.

Myasthenia gravis (MG) is a disease with muscle weakness due to impaired neuromuscular transmission, partly caused by autoantibodies against acetylcholine receptors on the motor endplate.

Lambert–Eaton syndrome is caused by autoantibodies against voltage-gated ion channels in neurons, which block fusion of vesicles at the motor endplate, causing muscle weakness.

Pemphigus is an autoimmune disease in which antibodies are directed against desmosomes (desmogleins 1 and 3) and disrupt adhesion between keratinocytes. This produces detachment of the epidermis and blistering. Pemphigus is strongly linked to DRB1*0401, a variant that presents a peptide of desmoglein very effectively.

Goodpasture's syndrome produces a type II reaction, in which autoantibodies directed against collagen type IV damage basement membranes in lung and kidney, leading to necrosis of the glomerulus and hemorrhage in the lungs.

TYPE III (IMMUNE-COMPLEX-MEDIATED) HYPERSENSITIVITY

Immune complexes are combinations of antigen and antibody, often with associated complement components.

Immune complex deposition Type III hypersensitivity results from the deposition of immune complexes in blood vessels and tissues. Sites of high blood pressure, filtration, or turbulence are particularly affected. Complexes can activate platelets (in humans) and basophils via Fc receptors, to release vasoactive amines, which cause endothelial retraction and increased vascular permeability, leading to complex deposition. Complexes also activate complement, releasing C3a and C5a; both activate basophils, and C5a is also chemotactic for neutrophils. Phagocytes that are unable to internalize the deposited complexes release granule contents and ROIs, causing local tissue damage.

Immune complex clearance In humans, circulating complexes are normally taken up by erythrocytes and carried to the liver, where they are transferred to, and degraded by, phagocytes. Factors that

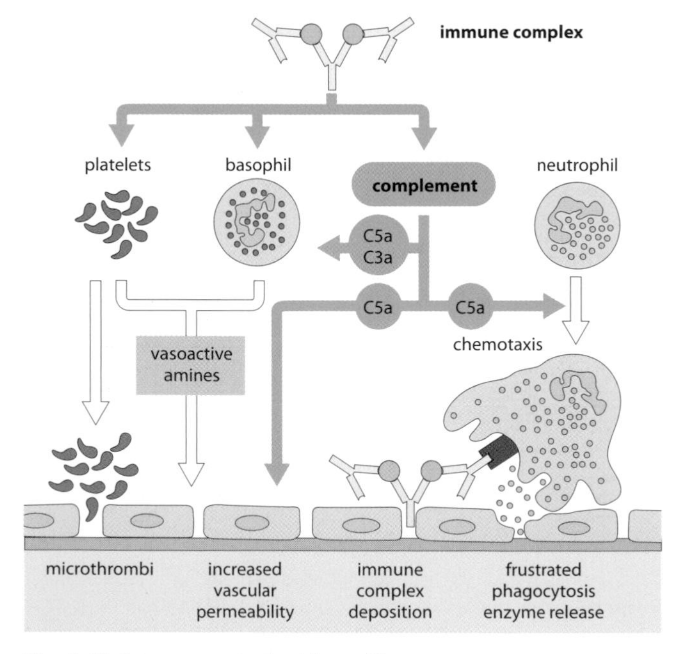

Fig. 4.12 Immune complex deposition.

affect clearance include: (1) size of complexes, (2) class and affinity of the antibody, (3) valency of the antigen, (4) the amount of complex. This last factor explains why immune-complex disease occurs in infections that release large amounts of antigen, and in autoimmune diseases where there is a quantity of autoantigen.

Immune-complex diseases result when excessive immune complex deposition occurs in particular organs. Figure 4.13 shows the symptoms of immune-complex disease due to autoimmunity (upper) and infection (lower).

Serum sickness is a type III reaction that occurs in individuals injected with foreign serum. Antibodies are made against the serum antigens and there is massive immune complex formation, producing arthritis and nephritis.

Arthus reaction is a skin reaction seen as an area of redness and swelling that is maximal 5–6 hours after intradermal injection of antigen. It is caused by IgG binding to injected antigen and triggering inflammation by type III mechanisms.

	circulating complexes	vasculitis	nephritis	arthritis	skin deposits
rheumatoid arthritis					
systemic lupus erythematosus (SLE)					
polyarteritis					
polymyositis dermatomyositis					
cutaneous vasculitis					
leprosy					
malaria					
trypanosomiasis					
bacterial endocarditis					
hepatitis					

Fig. 4.13 Immune-complex diseases: sites of deposition.

TYPE IV (DELAYED) HYPERSENSITIVITY (DTH)

Delayed hypersensitivity includes a number of reactions that are maximal more than 12 hours after challenge with antigen and are dependent on T cells rather than on antibody. The cells responsible are primarily $CD4^+$ T cells, and the reactions are of three main types, which may occur concomitantly or sequentially depending on the antigen and how long it persists.

Contact hypersensitivity produces an eczematous skin reaction in sensitized humans that is maximal 48 hours after contact with the allergen. The allergens may be large molecules or small haptens (such as nickel), which attach to the normal body proteins and modify them. Dermal dendritic cells and Langerhans cells pick up these antigens and transport them to local lymph nodes, where they are presented to T cells. On rechallenge with the allergen, sensitized T cells migrate into the skin, producing a reaction characterized by mononuclear cell infiltration, with edema and microvesicle formation in the epidermis. The dermis is usually infiltrated by an increased number of leukocytes. Keratinocytes have a key role in the development of the reaction, by the secretion of TNF-α, IL-1, GM-CSF, CXCL2, and CXCL10. They also contribute to the subsequent resolution of the reaction by production of IL-10 and TGF-β.

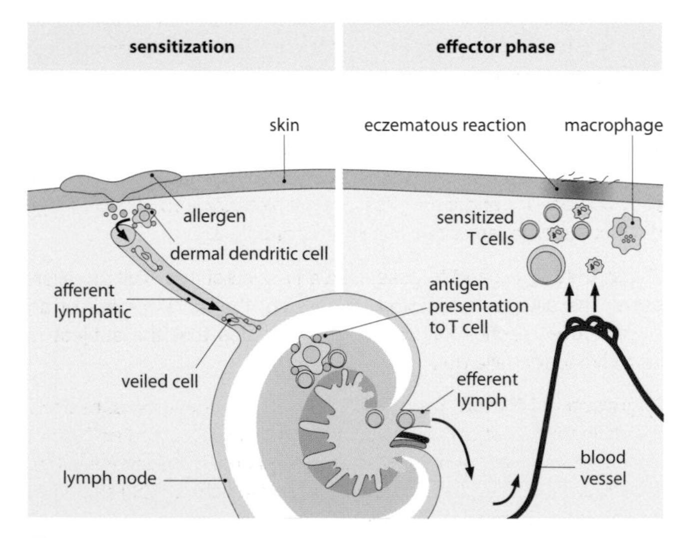

Fig. 4.14 Sensitization and effector phases of contact hypersensitivity.

Tuberculin-type hypersensitivity was originally a reaction produced by subcutaneous injection of tuberculin in patients with tuberculosis, who responded with fever and swelling at the injection site. The term now refers to the skin reaction induced by intradermal antigen that is maximal at 48 hours after challenge and consists of lymphocytes and mononuclear phagocytes. If the antigenic stimulus persists, a granulomatous reaction may develop. This type of reaction may be induced in sensitized subjects by several microbial and nonmicrobial antigens.

Granulomatous reactions develop where there is a persistent stimulus that macrophages cannot eliminate. Nonantigenic particles (such as talc) induce nonimmunological granulomas, whereas persistent pathogens such as mycobacteria and *Schistosoma* spp. induce immunological granulomas. Activated macrophages have been divided into M1 and M2 types, paralleling the TH1 and TH2 types of immune response. Granulomatous reactions induced by pathogens are of the TH1/M1 type, induced by IFN-γ. Such lesions consist of a palisade of epithelioid cells and macrophages surrounding the pathogen, which is in turn surrounded by a cuff of lymphocytes. Collagenous capsules may also develop around some pathogens, as a result of of fibroblast proliferation.

Epithelioid cells are large flattened cells with large amounts of endoplasmic reticulum, seen in granulomas. They are thought to be derived from macrophages, although they have fewer phagosomes than macrophages. Cytokine formation by these cells (for example, TNF-α) is important in the granulomatous reaction.

Giant cells are large multinucleate cells present in granulomas, which are derived from the fusion of macrophages and epithelioid cells. They may be induced by foreign bodies that cannot be readily phagocytosed. Langhans giant cells, which are a morphologically distinct subtype, are induced by IFN-γ and IL-3.

Patch testing is used to assess type IV contact sensitivity to allergens. The allergen is applied to the skin; the development of an eczematous reaction 48 hours later indicates that the subject is sensitive to that allergen.

Migration inhibition test (MIT) This assay detects sensitized T cells *in vitro*. Test cells are packed with monocytes and antigen in capillary tubes and then cultured on agar plates. If antigen-sensitive T cells are present, they release cytokines (such as MIF) that limit the migration of the monocytes.

5 Immunological Techniques

ANTIBODIES AND ANTIGENS

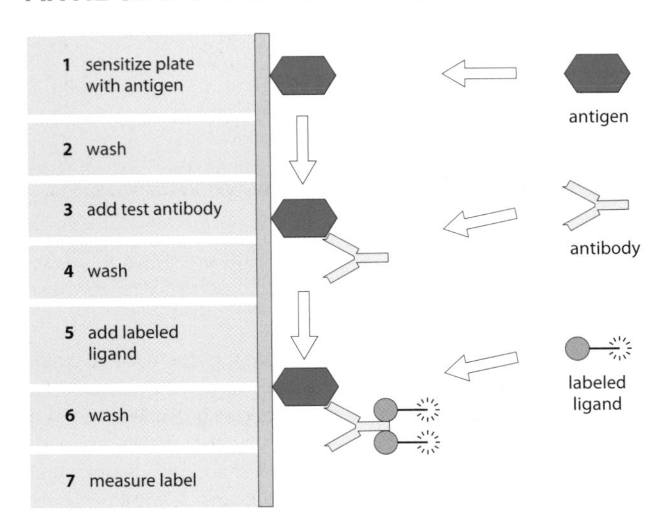

Fig. 5.1 Radioimmunoassay.

Radioimmunoassay (RIA) includes a variety of techniques that use radiolabeled reagents to detect antigen or antibody. Antibody may be detected using plates sensitized with antigen (Fig. 5.1). Test antibody is applied and this is detected by the addition of a radiolabeled ligand specific for that antibody. The amount of ligand bound to the plate is proportional to the amount of test antibody. RIA ligands are usually antibody molecules or protein A covalently bound to ^{125}I.

Protein A and **Protein G** are cell-wall components of staphylococci that bind specifically to IgG (Fc) of most species at a site between $C_{\gamma}2$ and $C_{\gamma}3$. Protein G binds a wider range of IgG subclasses than protein A.

Streptavidin/biotin reagents are used in many immunoassays (such as RIA and ELISA) to amplify detection and reduce background. Streptavidin binds biotin with very high affinity. For example, the antibody in Fig. 5.3 (ELISA) could be biotinylated, and this would then be detected with an enzyme coupled to streptavidin.

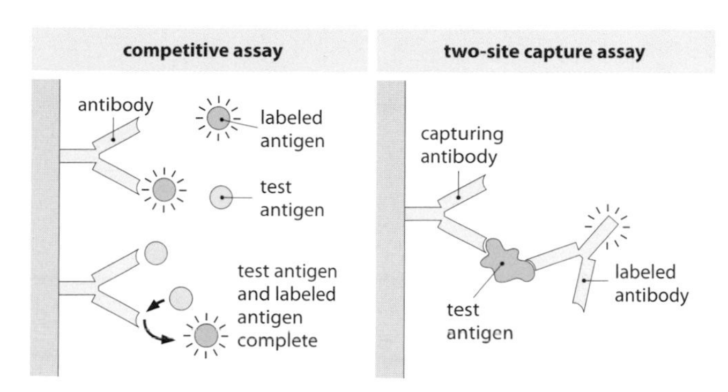

Fig. 5.2 Competition and sandwich (capture) immunoassays.

Radioallergosorbent test (RAST) is a specialized RIA to detect antigen-specific IgE. Antigen is covalently coupled to cellulose discs and specific IgE is detected with radiolabeled anti-IgE.

Competition radioimmunoassay is the classic RIA, used to quantitate antigens. Specific antibody is bound to a solid phase and a mixture of test (unlabeled) and labeled antigen is applied. Labeled and unlabeled antigens compete with each other for the antibodies' binding sites. The greater the amount of test antigen present, the less labeled antigen will bind to the antibody. Calibration curves using known quantities of unlabeled antigen are established. The technique is often used to assay hormones.

Radioimmunosorbent test (RIST) is a competition RIA used to detect IgE (the antigen in the assay), in which test IgE is competed with labeled IgE on plates sensitized with anti-IgE.

Sandwich (capture) immunoassays use antibody bound to the solid phase to capture molecules (antigens) from the test solution, which are then detected with a second labeled antibody. For example, solid-phase anti-IFN-γ captures IFN-γ from the test solution, which is detected with a second labeled antibody that binds a different site on the IFN-γ. Such assays can detect antigen at less than 1ng/ml and are often used to detect cytokines. For nonradioactive detection, an enzyme or a fluorescent tag can be coupled to the detection antibody (see ELISA and FIA overleaf).

Immunoradiometric assay (IRMA) is a test for antigen in which excess specific labeled antibody is added to the test antigen, which binds and neutralizes some of the antibody—free antibody is removed by solid-phase antigen, and so the residual radioactivity of the solution is proportional to the amount of test antigen.

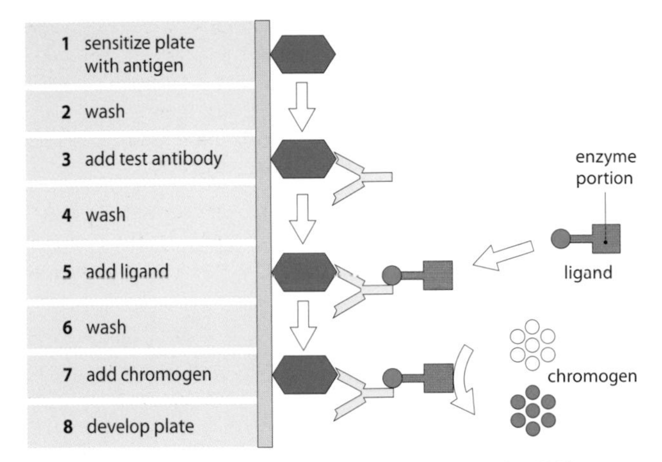

Fig. 5.3 Enzyme-linked immunosorbent assay (ELISA).

Enzyme-linked immunosorbent assay (ELISA) is used for detecting antibodies in ways analogous to RIA, but with the substitution of an enzyme for the radioactive isotope. For example, antigen is adsorbed on wells in a plate and test antibody is added that is detected using enzyme-labeled protein G (binds IgG). Enzymes such as peroxidase or phosphatase are often used. In the final stage a chromogenic substrate is added, which generates a colored end-product in the presence of the enzyme portion of the ligand. The optical density of the solution is measured after a defined period; this is proportional to the amount of enzyme, which in turn is related to the amount of test antibody. ELISA detection reagents, strictly speaking, are enzyme conjugates, but fluorescent or chemiluminescent tags can be substituted for the enzymatic detection system. Compared with RIA, ELISA has the advantage of stable reagents, but is usually less sensitive and less linear.

Nephelometry is an assay used to detect antigen or antibody by the formation of immune complexes. The complexes make the solution turbid, and this can be detected by light scatter.

Fluorescence immunoassay (FIA) is analogous to RIA, but substitutes fluorescent reagents for the radiolabeled material. The method has the advantage that fluorescent reagents may be detected instantaneously, but problems can arise with intrinsic fluorescence of the test material and with the availability of suitable reagents. Some fluorescent reagents respond differently when they are bound to antibody than when free, and this is the basis of a number of assays, for example:

Fluorescence quenching is the decrease in fluorescent light emitted by an antibody (or antigen) when it forms a complex. For example, it occurs when a hapten that absorbs radiation at 350 nm is bound to an antibody. Normally, antibody illuminated at 280 nm fluoresces at 350 nm, but if the hapten is bound, some of the fluorescence is absorbed (quenched).

Fluorescence enhancement is the increased fluorescence produced by some haptens when bound to antibody. The energy is absorbed by the antibody and emitted at a wavelength characteristic of the hapten.

Fluorescence polarization If polarized light is directed at a fluorescent molecule, it is absorbed and emitted shortly afterward, during which time the molecules move at random, so that the fluorescent emission shows reduced polarization. If, however, the fluorescent molecule is bound to an antibody, it has less rotational freedom and the initial polarization is retained when the fluorescent light is emitted.

Fluorescence (resonance) energy transfer is used to determine the proximity of two molecules on a cell surface. One is labeled with a fluoresceinated antibody (donor) and the other with a rhodaminated antibody (acceptor). The cells are illuminated at the donor's wavelength. If the two molecules are sufficiently close (less than 10 nm), energy is transferred between the fluorochromes and is detected at the acceptor's emission wavelength (red).

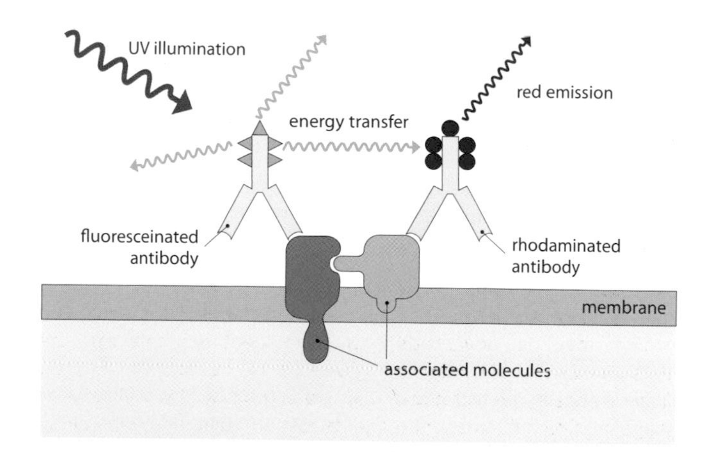

Fig. 5.4 Resonance energy transfer.

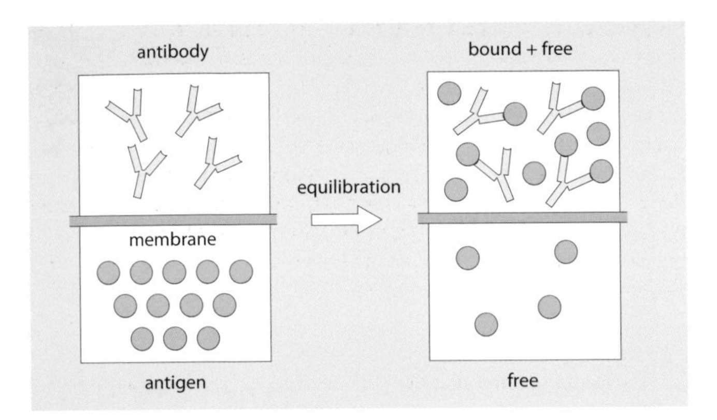

Fig. 5.5 Equilibrium dialysis.

Equilibrium dialysis is a method for determining antibody affinity in which a dialyzable antigen and the test antibody are placed in chambers on opposite sides of a membrane. The system is left until the concentration of free antigen is the same on either side of the membrane (equilibrium) and then the solutions are sampled. The average affinity (K_0) is defined as the reciprocal of the free antigen concentration when half of the antibody's combining sites are occupied; for IgG: Affinity, $K_0 = 1/[Ag_{free}]$.

Hemagglutination This term covers a number of techniques for detecting antibodies, based on the agglutination of red blood cells. The antigen may be either a red cell antigen or a molecule that has been chemically linked to the cell surface. For the test, the antibody is titrated in wells and the sensitized cells are added. If antibody against the red cell is present, the cells sink as a mat to the bottom of the well, but if it is absent they roll down along the sloping slides of the well to form a pellet.

Direct and **Indirect Coombs tests** are hemagglutination assays that detect antibodies against red cell antigens. The direct Coombs test identifies antibodies that can themselves cross-link the red cells. The indirect Coombs test detects antibodies that cannot cross-link the cells alone (for example, because there are too few antigens), by adding an anti-antibody that binds the first antibody.

Complement fixation test detects antibody or antigen. The test antibody is mixed with the antigen and a small amount of active complement. If antibody is present, complexes form and fix the complement. If there is any residual active complement it can be detected by the lysis of antibody-sensitized erythrocytes (EA).

Immunoblotting (Western blotting) is used to identify proteins that have been separated by gel electrophoresis (usually SDS polyacrylamide gel electrophoresis, SDS-PAGE) and then transferred to a membrane (blot). The blot is incubated with a primary antibody, which binds to the target antigen on the blot. The primary antibody is then detected with a secondary antibody conjugate coupled to (for example) an enzyme, fluorescent, radiolabeled, or chemiluminescent tag (cf. ELISA, RIA, and FIA). Primary antibodies for use in immunoblotting must be carefully selected so that they can recognize denatured antigens on the blot.

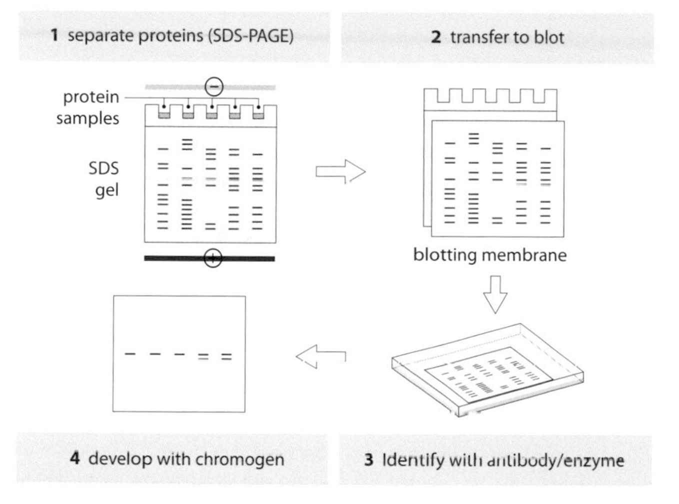

Fig. 5.6 Immunoblotting.

Immunoprecipitation is used for characterizing the antigen recognized by a monoclonal antibody, particularly if the antigen is denatured by immunoblotting. The antigen mixture is labeled (radiolabel, biotin, etc.) and precipitated in solution with the monoclonal antibody and co-precipitating agent (protein A, anti-IgG, etc.). The precipitate is then separated by SDS-PAGE and the labeled antigen is localized on the immunoblot.

Immuno-coprecipitation is used to detect whether two antigens are associated with each other. In the first step, the antigen mixture is immunoprecipitated with a primary antibody against antigen X. The precipitate is then analyzed by immunoblotting to see whether it contains antigen Y, using a primary antibody against Y. If Y is present, it indicates that it was associated with X.

Precipitin reactions When antigen and antibody react together near their equivalence point, they often form cross-linked precipitates. If the reaction occurs in a supporting medium such as an agar gel, the reactants form precipitin arcs, which may be used to identify or quantitate antigens and antibodies in complex mixtures. The methods include immunodiffusion (Ouchterlony technique), used to identify the relationship between antigens, and single radial immunodiffusion (Mancini technique), used to quantitate antigens. These techniques have largely been superseded because they are time-consuming and use large quantities of reagents.

Immunoabsorption is used specifically to remove particular antibodies from a solution, by the addition of a solid-phase antigen immunoabsorbent. Absorbents can include cells, chemically cross-linked antigen precipitates, and proteins coupled to solid supports.

Affinity chromatography is used to isolate pure antibodies. A column is prepared from antigen covalently coupled to an inert solid phase, such as cross-linked dextran beads. The antibody-containing solution is run into the column in a neutral buffer. Specific antibody binds to the antigen; unbound antibody and other proteins are washed through. The specific antibody is eluted using a buffer that dissociates the antigen–antibody bond; that is, high or low pH, or denaturing agents. By using antibody bound to the solid phase, the technique can be used to isolate antigens.

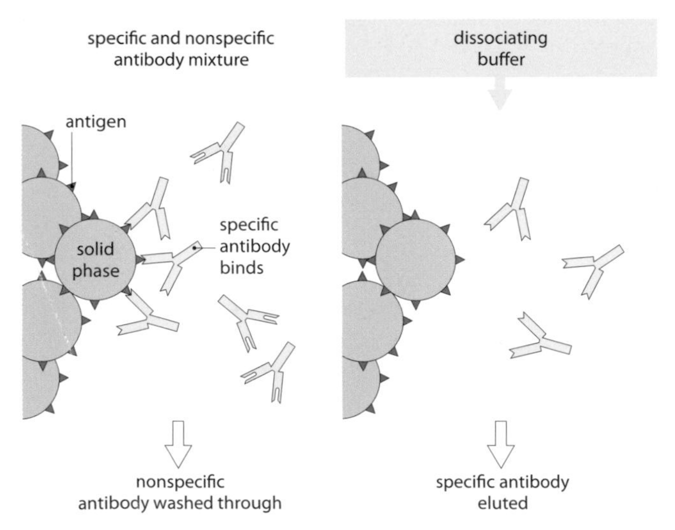

Fig. 5.7 Affinity chromatography.

Optical biosensors (surface plasmon resonance) are instruments that measure interactions between ligands in real time, based on the phenomenon of surface plasmon resonance. One of the reactants is immobilized on a chip coated with a film of gold, and the potential ligands are passed over it in the fluid phase while the chip is illuminated by polarized light. If ligands bind to the immobilized reactant, the optical properties of the chip are altered, thereby affecting the reflected light. Such instruments are useful for measuring the rate of association and dissociation of, for example, antibodies and antigens. They are also appropriate for detecting lower-affinity interactions between molecules, such as the binding of cell-adhesion molecules to integrins.

Electrophoretic mobility-shift assays (EMSA) are techniques to detect the association of two molecules, by running them (electrophoresis) through a nondenaturing, polyacrylamide gel. Bound molecules have different characteristics (size and/or charge) from unbound molecules and therefore their mobility will be shifted in the gel. The technique can be used to determine (for example) whether a transcription factor is associated with a DNA segment. If a complex of a transcription factor and a DNA probe is present, then the identity of the factor can be determined by adding antibody against it into the mixture (Fig. 5.8). In this example, free probe (lane1) moves fastest; probe bound to a nuclear protein (lane 2) is shifted. Addition of antibody against transcription factor (TF) A (lane 3) supershifts the band, indicating that the protein bound to the DNA probe is transcription factor A. Antibody against a different transcription factor, B (lane 4), has no effect.

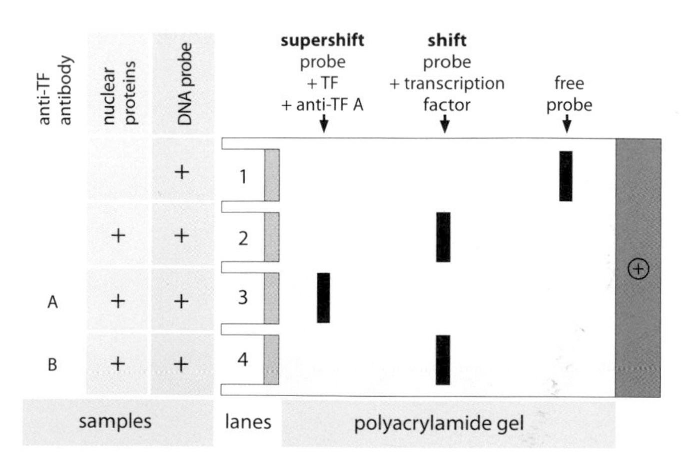

Fig. 5.8 Electrophoretic mobility-shift assays (EMSA).

Immunofluorescence is a general method for identifying antigens in tissue sections and on cells, or for identifying antibodies against them, as follows:

Direct immunofluorescence The antibody is covalently coupled to a fluorescent molecule such as fluorescein or rhodamine, which is then incubated with the cells or a frozen tissue section. (Some antibodies bind to wax-embedded sections, but not all.) The antibody is then visualized by observing the material under a microscope with incident UV illumination.

Indirect immunofluorescence In this technique the section is incubated with the test antibody, which is then visualized by the addition of a second-layer fluorescent anti-antibody. The amplification produced by the second antibody increases the sensitivity of the assay, and by using class- or subclass-specific reagents, particular antibody isotypes can be identified in the test serum. This technique is of great value for identifying antibodies against tissue antigens, as illustrated below, where antibodies against a pancreatic islet of Langerhans in diabetic serum were identified using indirect immunofluorescence on a frozen section of pancreas.

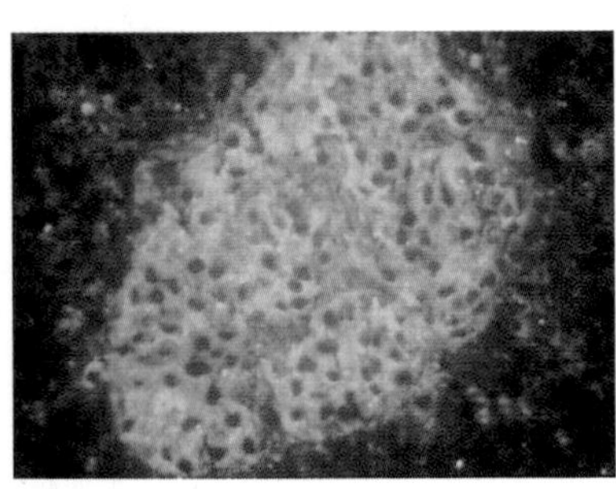

Fig. 5.9 Immunofluorescence: islet cell autoantibodies. Courtesy of B. Dean.

Capping occurs when antibodies bind and cross-link antigens on a live cell. The antigens aggregate at one pole of the cell and are then internalized by endocytosis.

Immunohistochemistry is similar to immunofluorescence, but enzyme-labeled conjugates and chromogens are substituted for the fluorescent conjugates, depositing an insoluble stain on the section. Sections are viewed with a light microscope.

Immunogold labeling is used to identify antigens by electron microscopy, using antibodies coupled to gold particles. By using gold particles of different sizes (5–25 nm), coupled to different antibodies, several antigens can be localized in the same section.

Flow cytometry is a technique that measures the characteristics of individual cells, including size, granularity, and fluorescence, as they pass through a flow cytometer in a stream of droplets. Cells may be stained with a set of different fluorescent antibodies to quantify the surface density of a number of different antigens on each cell. Populations of cells can then be identified according to their profile of surface molecules.

Fluorescence-activated cell sorter (FACS) is an instrument that carries out flow cytometry on a mixed population of cells. Basic instruments analyze the cells and can quantitate the proportions and phenotype of each subpopulation. A sorter is also able to separate the cells into different subpopulations so that they can be used in subsequent experiments. The parameters for the sorting (size, fluorescence intensity, etc.) are set by the operator. Figure 5.10 shows the basic arrangement of a FACS. The cell sample is carried in sheath fluid and split into droplets. Each drop containing a cell is illuminated by a laser; detectors identify side scatter (granularity), forward scatter (cell size), and fluorescence (specific markers). The data are fed to a computer, which controls electronic gates that steer the droplets into collection tubes to recover the isolated cell samples.

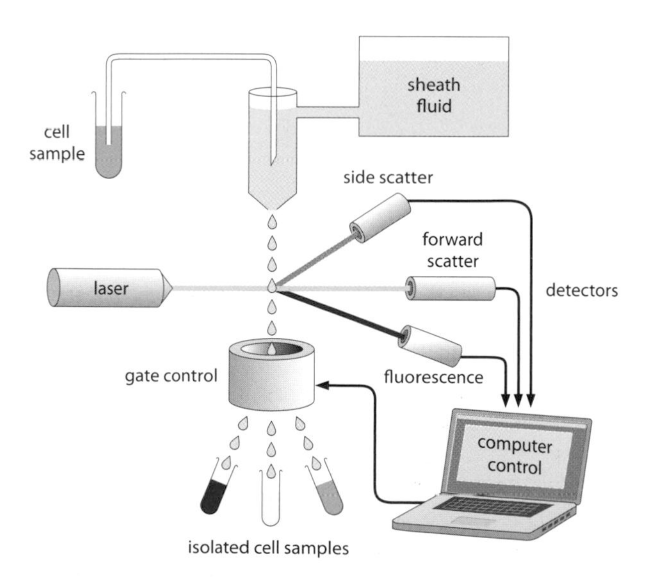

Fig. 5.10 Fluorescence-activated cell sorter (FACS).

CLONES AND CELL LINES

A clone is a group of cells derived from a single original cell; they are therefore genetically identical. A cell line is a group of cells grown in defined conditions from an initially heterogenous population. Only occasionally will such a line be monoclonal.

Immortalization describes the process by which a cell with a finite lifespan is genetically modified so that it can divide indefinitely.

Hybridomas are cells produced by the physical fusion of two different cells. Polyethylene glycol (PEG) and Sendai virus are often used to effect fusion. A hybridoma cell and its progeny contain some chromosomes from each fusion partner, although some of them will be lost. Hybridoma technology underlies the production of monoclonal antibodies.

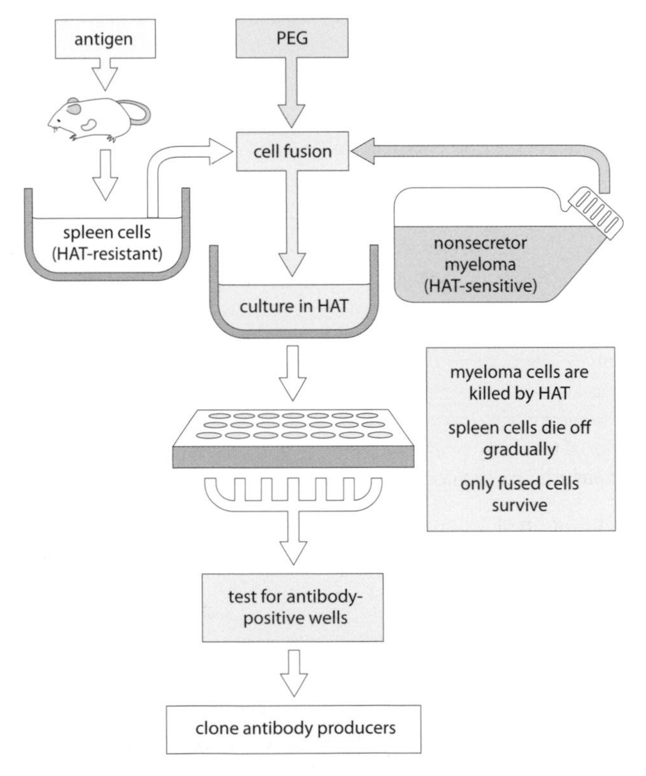

Fig. 5.11 Production of monoclonal antibodies.

Monoclonal antibodies are homogenous antibodies produced by a single clone. They are often made from hybridomas, which are prepared by fusing lymphocytes (for example, splenocytes) from an immunized mouse or rat, with a nonsecretor myeloma cell line using PEG (Fig. 5.11). The fusion mixture is plated out in a selection medium such as HAT. HAT contains hypoxanthine, aminopterin, and thymidine. Aminopterin blocks a metabolic pathway that can be bypassed if hypoxanthine and thymidine are present, but the myeloma lacks this bypass and consequently dies in HAT medium. Lymphocytes die naturally in culture after 1–2 weeks, but fused cells survive because they have the immortality of the myeloma and the metabolic bypass of the lymphocytes. Some of the fused cells secrete antibody, and supernatants are tested in a specific assay. Wells that produce the desired antibody are then cloned. Human B cells can be immortalized by transformation with Epstein–Barr virus. By comparison with polyclonal antisera, monoclonal antibodies are well defined, but not always more specific or of higher affinity.

Cloning is a process in which a cell population is successively diluted and set up in culture so that there are wells containing only one cell. The progeny of this cell are grown as a clone. Alternatively, the cultures may be grown in soft agar to prevent them spreading, and colonies are isolated by micromanipulation.

T-cell lines are produced by culturing a population of primed T cells in the presence of antigen and IL-2, which promotes proliferation of antigen-specific cells. The antigen must be presented to the T cells by APCs, usually macrophages or thymocytes, that have been treated to block their metabolism. The phenotype of the T cells can be modulated during production by the addition of other cytokines. For example, expanding T cells in the presence of IL-4 and corticosteroids favors the generation of TH2 cells, whereas the standard protocol favors TH1 cells. The production of antigen-specific T cells is measured by proliferation.

Proliferation of lymphocytes is usually measured by their uptake of radiolabeled metabolites required for DNA or RNA synthesis, such as ^{125}I-uridine deoxyribose or ^{3}H-thymidine. The uptake of these metabolites is measured by harvesting the cells on a cell harvester and counting the incorporated radioactivity.

Peptide:MHC complexes are used to identify clones of T cells that recognize a particular antigenic peptide. The complex consists of biotinylated MHC molecules (bound as tetramers to avidin) with the appropriate antigenic peptide. These MHC:antigen polymers have high avidity for the TCR and stimulate T cells very effectively.

ISOLATION OF CELLS

Ficoll gradients are used to isolate cells of different densities. In particular, they are used in the purification of lymphocytes. A diluted blood sample is layered onto the Ficoll and centrifuged. Because red blood cells and polymorphs are denser than Ficoll, they sediment to the bottom of the tube, whereas the lymphocytes and some macrophages remain at the interface. Lymphocyte populations may be further depleted of macrophages by adherence, or by letting the phagocytes take up iron filings and then removing them with a magnet.

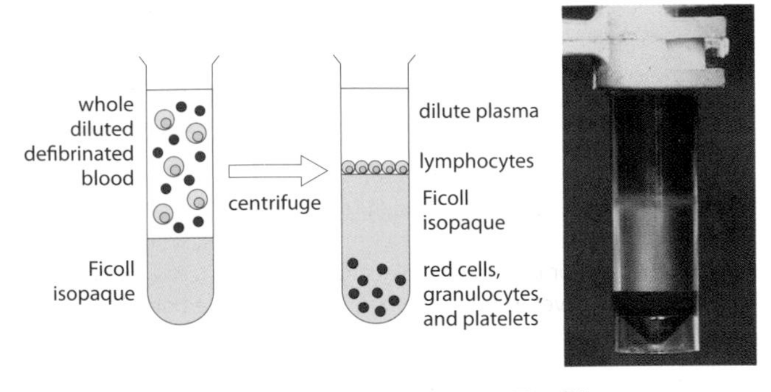

Fig. 5.12 Separation of lymphocytes on a Ficoll Isopaque gradient.

Adherence Macrophages have the property of adhering to plastic; they may be removed from cell suspensions by plating on plastic dishes, to which they adhere.

Panning uses plastic tissue-culture plates sensitized with antigen or antibody (cf. RIA). Mixtures of cells are incubated on the plate, and cells with receptors for the sensitizing agent bind to it. For example, cells with an antigen receptor will bind to an antigen-coated plate. Alternatively, cells with a particular surface marker are depleted by attachment to a plate coated with antibody against that marker. The technique is mostly used to deplete cells of a specific subpopulation. However, the bound cells can sometimes be recovered by chilling or by digesting the plate with enzyme. One limitation is that the process of binding to the plate may cross-link the surface receptors on the cells and activate them.

Antibody/Complement depletion Specific cell populations can be removed from a mixture by lysis with antibody and complement.

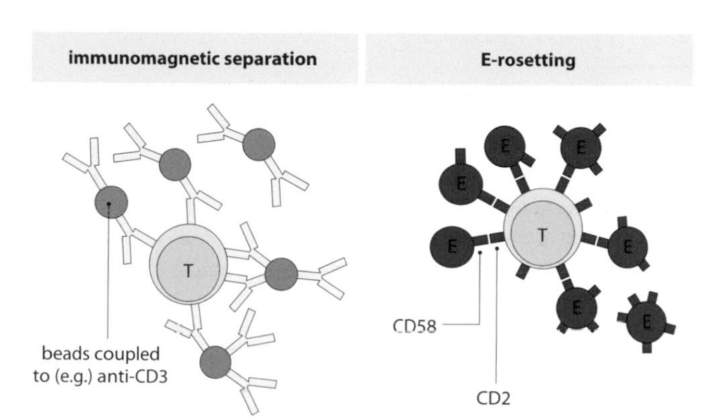

Fig. 5.13 T-cell separation by immunomagnetic beads or rosetting.

Immunomagnetic beads are an efficient way of isolating cell populations in bulk. The cells are mixed with magnetic beads coupled to a particular antibody (for example, anti-CD4). They may then be rapidly removed or isolated by placing the tube in a magnetic field. Cells can be recovered by detaching them from the beads.

Rosetting is a method of isolating cells by allowing them to associate with red blood cells. Lymphocytes become surrounded (rosette) with red cells and may then be isolated by sedimentation through Ficoll gradients. Human T cells have receptors for sheep erythrocytes (E) and so may be isolated by mixing with sheep cells and separating the rosettes. Cells that have Fc receptors for IgG or IgM can be isolated by mixing with red cells sensitized with antibody of the appropriate class; the antibody cross-links the red cell to the Fc receptor. Red cells coated with specific antibodies against surface markers of different leukocyte populations have also been developed. In all cases the cells are recovered by density-dependent centrifugation.

Antigen suicide is used to deplete those cells of a population that bind a particular antigen, by supplying them with a highly radioactive antigen, which is taken up and kills the cell. A modification of this technique to kill proliferating cells is to add bromodeoxyuridine, which they incorporate into their DNA. Illumination with UV radiation activates the metabolite to kill the cells.

Percoll gradients Percoll is a medium that can be used to form density gradients by ultracentrifugation. Cells are layered onto the top of the gradient and spun. Different cell populations settle at different positions (bands) depending on their density.

CELLULAR FUNCTIONS

Plaque-forming cells (PFCs) are antibody-secreting cells measured in an assay where each cell produces a clear zone of lysis (plaque) in a layer of antigen-sensitized red blood cells. In the assay, lymphocytes are mixed with the sensitized red cells and placed in a chamber slide. Antibodies released from the B cells bind to the surrounding red cells, which are lysed by the addition of complement. The assay can be modified to distinguish IgM-producing B cells from IgG producers. Total antibody-producing cells (not just the antigen-specific ones) are measured in the reverse plaque assay, in which the red cells are sensitized with anti-Ig or protein G, which binds all released IgG antibody.

ELISPOT assays are enzyme immunoassays used to quantify antigen-specific cells. Antibody-forming cells are detected by overlaying lymphocytes on a plate sensitized with the specific antigen. Specific antibody binds to the antigen immediately around the cells secreting it. This can be detected by an enzyme immunoassay, which deposits an insoluble chromogen around the secreting cell (Fig. 5.14), appearing as a small colored spot. The technique is also used to detect the numbers of cells secreting a particular cytokine. For example, active TH1 cells can be detected by overlaying them on plates sensitized with antibody against IFN-γ to capture secreted IFN-γ. The spot of cytokine is detected using an antibody against a different epitope of the IFN-γ.

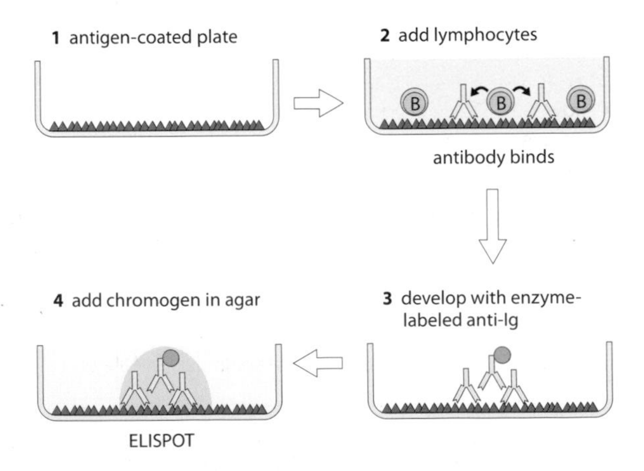

Fig. 5.14 ELISPOT assay to detect Ig-secreting B cells.

Chromium release (cytotoxicity) assay is used to measure the activity of cytotoxic cells. The target cells are first mixed with radioactive ^{51}Cr, which is taken up by viable cells. These are then incubated with the test leukocytes. If the test cells damage the targets, released ^{51}Cr can be measured in the supernatant.

Trypan blue is a stain that can be used to assess cell viability. Dead cells take up this dye, which is used in cytotoxicity assays.

MTT assay is a colorimetric assay for cellular metabolic activity, in which the dye MTT is reduced by NADPH to produce a purple compound; staining density reflects energy generation by the cell.

NBT (Nitroblue tetrazolium) reduction is a standard test for the oxidative burst of neutrophils; it produces a deep blue end-product from the colorless chromogen NBT.

Adhesion assays are used to detect interactions between different cell populations, particularly leukocytes and endothelium. The simplest method is to examine the co-culture of leukocytes on top of the endothelium (phase bright) and the migrated cells beneath the endothelium (phase dark). To quantify the migrated cells, the leukocytes can be prelabeled with cell trackers or a radioactive tracer (such as ^{51}Cr). Adhesion *in situ* is measured by the Stamper–Woodruff assay, in which leukocytes are overlaid on frozen tissue sections containing the blood vessels under investigation. Sections are examined under the microscope for evidence of leukocyte adhesion to the vessels. This technique was first used to identify the function of high endothelial venules (HEV) in lymph nodes. The molecules involved in adhesion have been identified both *in situ* and *in vitro* by adhesion blocking with specific antibodies (such as anti-VLA-4). More sophisticated adhesion assays are carried out under flow conditions *in vitro* that mimic flow and shear *in vivo*. Adhesion assays can also be carried out *in vivo*, by the use of fluorescent-labeled cells and intravital microscopy.

Photobleaching recovery is a technique used to measure the lateral mobility of molecules on a cell membrane. A molecule is labeled by a fluorescent antibody and a spot of membrane is bleached by extended illumination with UV radiation. The rate at which unbleached labeled molecules reenter the bleached area after the UV is shut off is a measure of molecular mobility.

***In situ* hybridization** is a molecular biological technique to detect expression of proteins (for example, cytokines). Tissue sections are hybridized with labeled cDNA and the cellular localization of the target mRNA is determined by microscopy.